MATHEMATICS

WITH NUMBERS IN COLOUR

BOOK VI

APPLIED MATHEMATICS

IN THIS SERIES

OTHER BOOKS ON THE CUISENAIRE-GATTEGNO APPROACH

Talks for Primary School Teachers   by Madeleine Goutard.

A Teacher's Introduction   by C. Gattegno.

MATHEMATICS
WITH NUMBERS
IN COLOUR

by

C. GATTEGNO

BOOK VI

EDUCATIONAL EXPLORERS LIMITED

READING

First Published 1961
Reprinted 1962
Revised and enlarged edition 1964
Reprinted 1968
© *C. Gattegno 1961–1964*

ISBN 978-0-87825-350-0

The original Cuisenaire material described in this book can be obtained in Great Britain from
THE
CUISENAIRE COMPANY LTD.
40 Silver Street, Reading, Berks.

Printed in Great Britain by
Lamport Gilbert Printers Ltd,
Reading, England

CONTENTS

I

LENGTH, AREA AND VOLUME

PART I

LENGTH, AREA AND VOLUME

The centimetre and its multiples

1. The rods we use have been so cut that their lengths are multiples of the *centimetre*, and their *cross-sections* equal, each being a square whose side is one centimetre long, i.e., one square centimetre.

The volume of the white rod is one *cubic centimetre*.

Give the length of all the rods in centimetres.

the red	the pink
the orange	the yellow
the tan	the black
the blue	the dark green
the light green	

2. Which rods would you choose to make the following lengths?

27 centimetres (27 cm.)	92 cm.
19 cm.	77 cm.
36 cm.	

The length of the orange rod is also called a *decimetre* or dm. Express in two ways the following lengths:

Seven orange rods end to end.

Three orange rods and two light green ones end to end.

Six black rods end to end.

Nine tan rods and two pink rods end to end.

3. The length of ten orange rods end to end is also called a *metre* or m.

Write a metre as dm. and then as cm.

What is the length in m., dm. and cm., of the following:

Eleven orange rods end to end;
Fifteen orange rods end to end;
Eighteen orange rods end to end;
Twelve orange rods and a red one end to end;
Twenty-two orange rods and one black end to end.

How would you make the following lengths:

1 m. 3 dm. 4 cm.	4 m. 7 dm. 2 cm.	3 m. 5 dm.
12 dm. 6 cm.	178 cm.	3 m. 4 cm.
9 m. 9 dm. 9 cm.	9 dm. 9 cm.	7 m. 5 dm.

Conversions

4.   Can you express one cm. and one dm. as fractions of a metre?

as decimal fractions?

Express the following lengths as a decimal of a metre:

132 cm.	15 dm.	19 cm.
30 dm.	12 dm. 8 cm.	731 cm.
100 cm.	0·1 dm.	7·2 dm.
3 m. 1 dm. 7 cm.	3 cm.	5 m. 7 dm. 9 cm.

Which of the following pairs is longer:

0·7 m. *or* 6 dm. 3 cm.	$\frac{3}{10}$ dm. *or* 0·03 m.
1·02 m. *or* 1 m. $\frac{3}{10}$ dm.	11 cm. *or* 1 dm. 1 cm.
3·2 m. *or* 321 cm.	1·1 dm. *or* 0·011 m.

Find the total length of the following trains of rods, and express the results in as many ways as you can:

30 cm. *and* 70 cm.

21 cm. *and* 8·8 dm.

3

1 m. *and* 171 cm.

12 m. 2 dm. *and* 1 m. 9 dm.

41 cm. *and* 64 cm.

3 dm. 2 cm. *and* 5 dm. 9 cm.

2 m. *and* 3 dm. 2 cm.

1 m. 3 dm. 5 cm. *and* 1 m. 7 dm. 6 cm.

Applications

5. From a piece of material 3 m. 4 dm. 7 cm. long a piece 1 m. 6 dm. 9 cm. long is cut. What is left?

From a piece of lino 2 m. 7 dm. long a piece 2 m. 5 cm. long is cut. How much is left?

How many pieces 1 m. 3 dm. 5 cm. long can you cut from an 8 m. long piece of material? How much is left?

How much lino must we buy to cover a floor 4 m. 75 cm. long and 4 m. wide with lino 1 m. wide?

Wallpaper 80 cm. wide is sold in rolls 15 m. long. What is the height of a square room which is 4 m. wide if, to be papered to the ceiling, it requires 4 rolls? (Doors and windows are not taken into account.)

Find the perimeter of rectangles whose sides are:

12 m. 7 dm. 5 cm. and 3 m. 6 dm.

1 m. 4 cm. and 72 cm.

3·14 cm. and 1·29 cm.

$\frac{3}{4}$ m. and $\frac{7}{10}$ m.

$1\frac{1}{2}$ m. and 0·38 cm.

Multiples and submultiples

6. When we measure lengths that are either much smaller or much bigger than those we can make with rods, we must use other units.

The *millimetre* is one tenth of a centimetre. You can see these marked on a 30 cm. ruler. Smaller units are called

4

tenths and hundredths of a millimetre; they are used in very precise work in factories using hard metals.

The thousandth of a millimetre is the *micron* and is used to measure the linear dimensions of bacteria and germs. Smaller units are called tenths, hundredths and thousandths of a micron, and are used to measure the linear dimensions of viruses, of components of cells, and of large molecules. Some of these also serve to measure the wavelengths of some radiations. In Parts IV and V of this book we shall meet them again.

The *kilometre* is a unit used to measure distances on roads and between towns, and lengths of large rolls of wire. The kilometre, or km. is equal to 1000 metres end to end.

More rarely, the units equal to 10 and 100 metres are used: they are called the *dekametre* and the *hectometre*. We shall only mention them for you to know that you have met them. Usually, distances are given as a decimal of the kilometre if they are bigger than 999 m.; otherwise metres are used.

What are the following distances in metres:

3·75 km.   9·700 km.   4·444 km.

Write the following as a decimal of a kilometre:

9732 m.   4810 m. (height of Mont Blanc)   883 m.

(Note: The heights of mountains are always given in metres even if above one thousand; so are athletic racing distances, except the Marathon.)

Applications

7. If two places are 1500 m. apart, are they also 1½ km. apart?

A roll of wire has 1875 m. of wire. How many rolls will give us 19 km. of wire?

The primary of a transformer takes 5000 coils, 2·2 cm. long on the average. Its secondary takes 20,000 coils, each 4·3 cm. long. Calculate the length of wire needed for each part of the transformer. Give your answer in metres and kilometres.

5

How much electric wire do you need to wire a hangar of which you choose the dimensions and the number and position of the points and bulbs?

Errors and approximations

8. If, with each of the 291 rods in your set, the saw has cut $\frac{1}{4}$ millimetre (mm.) less than the proper length, how much shorter than 1 m. will the following lengths be?

> ten orange rods end to end,
>
> eleven blue rods end to end,
>
> twelve tan rods end to end.

What would be the length of 50 white rods end to end? of 50 red ones? of 33 light green ones?

Would 3 light green ones still equal to a blue?

How will 4 light green rods end to end compare with an orange and red end to end?

Which choice of rods representing 84 will produce the least *error*? What is the error if you only use white ones, red ones, etc.?

9. It is almost impossible to cut a rod so that it has exactly the required length. There is always an error. But some rods are slightly bigger and some smaller than they should be. Take any ten rods of any colour; compare their length and their width respectively. Can you find whether the errors *compensate* if you take any ten *at random* and put them side by side and measure the width with an orange rod? The width of nine side by side with a blue one? Of eight with a tan one, etc.?

What do you observe?

Devise other means of comparing your rods to judge their accuracy.

Absolute and relative errors

10. An error of $\frac{1}{4}$ mm. or 0·25 mm. affects rods of different lengths differently. The error in each is called the *absolute*

6

error, and is independent of the length in which it occurs. But the *relative error* changes according to the length. Thus, an error of ¼ mm. in a white rod is greater in relation to its length than the same error in an orange one, as we should soon see if we measured ten such white rods against one such orange rod. How much shorter than the orange would the white line be, if all the white rods were ¼ mm. too small?

We can express relative error in percentage form:

white rod: $\frac{0.25}{10}$ error $= \frac{2.5}{100}$ or 2·5% error

orange rod: $\frac{0.25}{100}$ error or 0·25% error.

The percentage error in the latter is $\frac{1}{10}$ that to be found in the former.

Now, put any two rods end to end. What is the error in that length if, in each rod, the absolute error is ¼ mm.?

If one is red and the other orange, find whether the relative errors in each add up to form the relative error in the total length?

Is $\frac{0.25}{20} + \frac{0.25}{100}$ equal to $\frac{0.50}{120}$? Is the new relative error always smaller than the sum of the relative errors?

11. If you subtract one length from another, each such length containing the same error of ¼ mm., what can you say about the error in the difference?

If you do not know whether your rods have been cut slightly longer or slightly shorter than they should be, but you suspect that each is liable to be up to ¼ mm. too long or too short, can you say between what limits of length a yellow, a red, and an orange rod put end to end will be?

Let *r* stand for the red rod, *y* for the yellow, *b* for the black, and *p* for the pink, so that, for example, 2*r*, 5*y*, 3*b*, 8*p* represent two red, five yellow, three black, eight pink rods respectively.

Find between what lengths the following arrangements of rods will fall if a possible error of ½ mm. occurs in each rod, but you do not know whether by excess or defect:

7

$$y+2b+3p+r$$
$$3r+4b+2y+5p$$
$$5b+4y+7p+9r$$
$$5y-3r+2p-b$$
$$2y+b-2\ (r+p)$$

Volume; the cubic centimetre

12. The white rod is a cube and its volume is one *cubic centimetre* (c.c. or cm.³).

Two white rods end to end have the same volume as one red rod.

Give the volumes in c.c. of one rod of each colour in your set.

With your rods you can make various bodies and find their respective volumes. For instance, four black rods can be put end to end to give a long cuboid 28 cm. long; or, taken two by two end to end, they form a cuboid with 2 square cm. base and 14 cm. long; or if the four are put side by side, they produce two different cuboids according to whether the base is a square 2 by 2 or a rectangle 1 by 4. The volume of each of these bodies is 28 c.c.

Take rods of the same colour, make cuboids with them, and find their volume. If the number of rods is odd, you can produce fewer cuboids than when it is even, but you can still· calculate the volume of each body you form.

Form a series of cubes using rods of one colour, starting with a cube with sides equal to 1 cm., then 2 cm. and so on up to a cube with sides equal to 10 cm.

Find in c.c. the volume of each.

Multiples

The volume of the cube on the orange rod is called the *cubic decimetre*.

How many c.c. are there in a cubic decimetre?

What is one c.c. of a dm.³?

Give the value of the following volumes:

10 c.c. as a fraction of dm.³;   100 c.c. as a fraction of a dm.³

30 „	„	„	125 „	„	„
56 „	„	„	250 „	„	„
59 „	„	„	373 „	„	„
66 „	„	„	492 „	„	„

Write the following as a decimal of a dm.³:

1 dm.³ 150 cm.³   4 dm.³ 35 cm.³   50 dm.³ 170 cm.³

How many c.c. are there in:

4 dm.³ 400 cm.³   5 dm.³ 43 cm.³   7 dm.³ 6 cm.³

13.   To measure large volumes such as tanks and reservoirs, the c.c. and dm.³ are too small. So the cube whose side is a metre is used; it is called the *cubic metre* or m³.

How many cubic decimetres of water do you need to fill a tank of one m.³?

How many cubic centimetres do you need for the same purpose?

What is a dm.³ of a m.³?

What is a c.c. of a m.³?

What is the volume of a cube with sides 50 cm.? or 25 cm.? Compare each with the m.³ and with the dm.³.

Express as a decimal of a m.³ the following volumes:

1 c.c.   7 c.c.   17 c.c.   50 c.c.   150 c.c.   1606 c.c.

21,732 c.c.   500,000 c.c.   1,800,000 c.c.   1 dm.³

3 dm.³ 80 cm.³   21 dm.³ 134 cm.³   40 dm.³ 45 cm.³

102 dm.³   327 dm.³ 100 cm.³   592 dm.³ 783 cm.³

7 cm.³ 1 m.³ 3 dm.³   2 m.³ 34 dm.³ 55 cm.³.

14.   To measure geological deposits or volumes of planets, the cube whose side is one kilometre is a more useful unit.

Can you readily grasp the idea of a cubic kilometre?

It may help if you place a white rod on the table and, near it, a cube made of wire whose side is one metre, and compare the two volumes. The hm.^{3} (or $100 \times 100 \times 100$ m.^{3}) will compare with the m.^{3} as this does with the cm.^{3}. Can you think of the relationship of the white rod to the km.^{3}?

Volumes of various bodies

15. The bodies whose volumes can be found easily are those which can be represented as cubes placed together in such a way that there is no space left between them and no overlapping.

With white cubes we can form a great variety of bodies which are not prisms, yet we can always find their volume by merely counting the cubes. Because the rods have been cut as multiples of the white one, they can be freely interchanged and, whatever shaped body we construct with any combination of rods we choose, its volume can be obtained by simple addition.

Not all natural bodies are formed in this way, yet we may need to know their volume. Two obvious variations are:

    (i)   bodies whose shapes remind us of those made of rods but which are formed by using some unit other than the centimetre, such as a fraction of the centimetre ($\frac{1}{n}$ cm.) or the inch or a fraction of an inch;

    (ii)   bodies with curved sides such as cylinders, spheres and the like.

In the case of the unit which is a fraction of a centimetre, we can, by taking a unit equal to $\frac{1}{n}$ cm., calculate the number of units of length in the sides of the figure, and the number of cubic units in the volume. For example, if we use $\frac{1}{3}$ cm. as the unit, the lengths will be represented by a number 3 times greater than when the cm. is used, and the volume by a number 27 times greater. Where the unit chosen is not related to the cm. as a fraction, we need to invent new methods of which we shall now say something.

10

Approximations

16. If we cut a rectangular prism out of a piece of wood without using any particular measure, it is most unlikely that its sides will be an exact number of cm. Accordingly, if we try to construct a body equal to it, using the white rods, its dimensions will be either bigger or smaller than the prism we made and its volume will, likewise, contain either more or fewer cm.^{3} than the exact number we wish to find.

If we could subdivide the cm.^{3} into small parts we could clearly make a body that would be closer in dimensions and volume to the prism we cut out. For example, if we had rods measuring 1 mm.^{3} (cubic millimetre) which is $\frac{1}{1000}$ of a cm.^{3} each, our chances of making with them a body equal to the prism would be increased and it would, in any case, be a much closer copy than one made of white rods.

This is so obvious that we need not press the point. But, even with mm.^{3}, it is unlikely that we could construct a body exactly equal to the original one. Both the bodies we constructed would be *approximations* to the original in dimensions and volume, and there would be an error in either case which would be greater with the white cubes than with the mm.^{3} cubes, because the latter would fill a space which remained unfilled when the former were used.

Every body has volume, but the value of that volume cannot always be found exactly by piling up cubes of our choice, such as cm.^{3} or mm.^{3}; but the value we can obtain in that way is 'near enough', if the cubes we choose are appropriate to the job we wish to do. (For the meaning of the expression, 'near enough', see Book V, Part V).

Since it is always possible to approximate more closely, (whether we obtain more or less than the exact answer) by decreasing the size of the cubes we use, we shall call the volume of a body, to a certain approximation, the volume of another body made of piled cubes where the difference in volume between the two is less than whatever limit we have set.

If a body has a volume larger than 135 cm.^{3} and smaller than 156 cm.^{3}, these numbers can be replaced by, let us say, 142,640 mm.^{3} and 147,320 mm.^{3} (which shows that the

11

substitution of mm.³ for cm.³ is not a mere transformation of the first two numbers by the relationship 1 cm.³=1,000 mm.³.) Clearly, we have a closer approximation with a volume that is given as lying between 142,640 mm.³ and 147,320 mm.³ than with one said to lie between 135,000 mm.³ and 156,000 mm.³.

So, every time we reduce the size of the unit we obtain a closer approximation.

The problem of finding the volume of bodies with curved sides is discussed in a later section when we consider capacity. Meanwhile, we shall look at what we have been doing from another point of view.

Areas and approximation of areas

17. This question of the approximation of volumes is better understood if we start with areas of plane figures.

If we place a few rods on a table, making a flat pattern, we can easily calculate, in square cm., the area of the top surface. It would be the number of white rods we need to cover that surface.

As long as we place white rods side by side, we can cover areas made of cm. rods and find their value.

For larger figures, it would be wearisome to use the small unit, and bigger ones can be used, the names of which are:

1 dm.², or square decimetre, which is equal to 10^2 or 100 cm.²,

1 m.², or square metre, which equals 100 dm.²=10,000 cm.².

For farm land or fields, the units are:

100 m.² or *are*

1000 m.² or *hectare*.

For geographical purposes, the square kilometre or km.², which is equal to $(1000 \text{ m.})^2$ or 1,000,000 m.², is preferred.

Now, if we draw any rectangle on a piece of paper the size of this page, it would be most exceptional if each side were equal to a whole number of cm. To measure it in cm. we should use decimal places to the number required to give the accuracy we wished to achieve.

Draw any rectangle and measure its sides in cm. and in mm. The one we decide upon has, we shall say, 12·52 and 6·73 cm. or 125·2 mm. and 67·3 mm. for the lengths of its sides. We can place 12 white rods along one side and 6 along the other side, but the 72 white rods do not quite cover the area of the rectangle. If we add one more row of white rods along each of the two sides, and then complete the white rectangle, we find it is bigger than the one we drew.

So the area A of the rectangle we drew is such that 72 cm.^{2} is smaller than A, while 91 cm.^{2} is bigger, or

$$72 \text{ cm.}^2 < A < 91 \text{ cm.}^2$$

If we now measure the rectangle using mm. paper, we see that we can put 125 lines of 67 squares each inside the rectangle, or 125×67 square millimetres, while the rectangle is contained in one made of 126×68 sq. mm. So A is such that

$$8,375 \text{ mm.}^2 < A < 8,568 \text{ mm.}^2$$

If we convert 72 cm.^{2} into mm.^{2}, we obtain 7,200, whereas 91 cm.^{2} gives 9,100. The numbers we found with the mm. paper are different from these and are closer to each other. The difference between the two numbers containing A is, in the first case 19 cm.^{2} or 1,900 mm.^{2}, while in the second it is only 193 mm.^{2}, which is about $\frac{1}{10}$ of the other.

So, when the unit is smaller, the value we obtain for each rectangle (the one inside and the one outside) is 'better', i.e. nearer to the value of A.

If we were able to use paper marked in $\frac{1}{10}$ mm. and counted the squares that form the two rectangles just inside and just outside the one drawn upon it, we should obtain a better value still.

Rectangles drawn *at random* have an area that cannot be an exact number of cm.^{2} or mm.^{2} or even of smaller units. So we accept a value to a certain approximation of 1, 2, 3 . . . decimals in any given unit.

Only exceptional rectangles have an exactly calculable area.

Figures reducible to rectangles

18. Figures other than rectangles have areas that can be calculated, but they present us with problems that must be solved before we can use the method we have discussed. This method, being based upon the white rod which is rectangular, can be used only when these non-rectangular figures can be reduced to rectangular ones without alteration of area. We can then use our method as before. The means by which we make the necessary rectangular areas can be illustrated by diagrams.

(a) *Parallelograms*

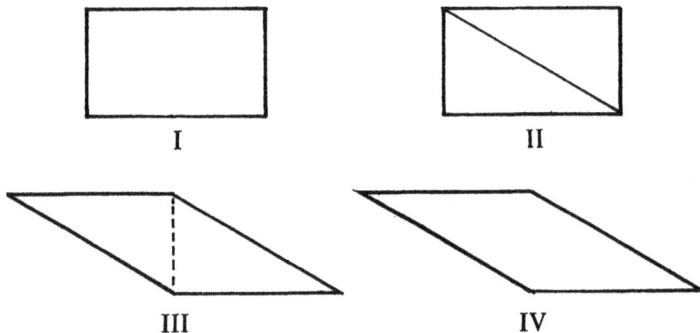

| I | II |

| III | IV |

If we begin with a rectangle and draw a diagonal, we see that the sections can be rearranged, as in the third figure. This is a parallelogram, as can be seen even more clearly when, as in the fourth figure, the line marking the join is erased. We can carry out this transformation with paper and scissors if we wish, and this makes it even simpler to follow.

Thus, we know that if we wished to find the area of the parallelogram in the fourth figure we could, by reversing the procedure, form the rectangle in the first figure, which would enable us to use our method to find the area.

But can *any* parallelogram be replaced by a rectangle in this way? The answer is obvious from the figures below.

By cutting off the right-angled triangle at one end and replacing it at the other we construct a rectangle of equal area. So it is always possible.

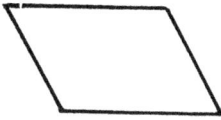

I

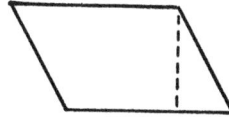

II

III

IV

(b) *Triangles*

Every triangle can be replaced by half a rectangle as we see if we study this figure:

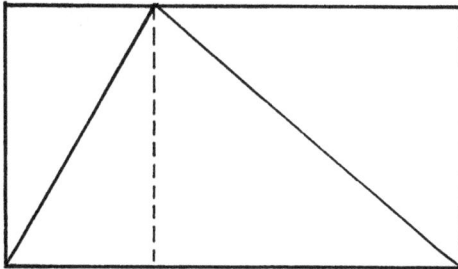

The perpendicular we have drawn gives us two rectangles, each of which is divided by a diagonal. The area of the triangle that falls within each rectangle represents half the area of that rectangle.

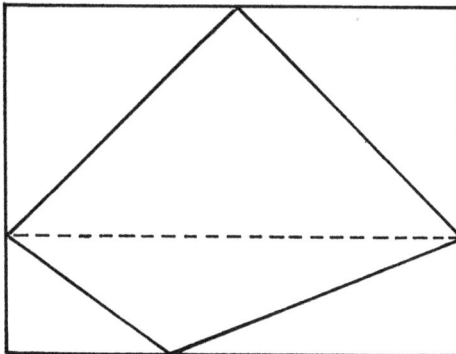

15

(c) *Quadrilaterals*

Any quadrilateral can be replaced by half a rectangle, as we see if we compare the figure above with the one in (b).

(d) *Polygons*

Polygons can be divided in several ways into non-overlapping triangles, the areas of which we know how to calculate.

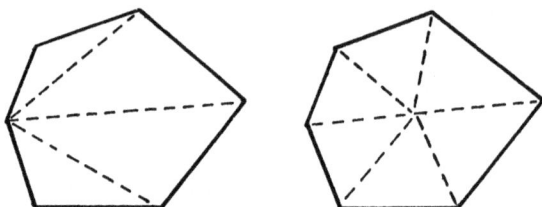

So, if we can measure the area of a rectangle in square units, we can also find the area of any quadrilateral (including parallelograms, rhombi and trapezia) or any triangle. And because we can find the area of triangles we can find the area of polygons.

Areas of more complex figures can only be found by methods of approximation. A circle, for instance, with radius r contains a square with area $2r^2$ and is contained in a square with area $4r^2$; it is, thus, about $3r^2$. Later on, we shall learn that the area of a circle is slightly more than $3r^2$, which is written as πr^2 (and read as *pi r squared*). π is an important number that crops up in many problems, and its value, to two decimal places, is 3·14 and, to four decimal places, is 3·1416. You will learn how to find it one day.

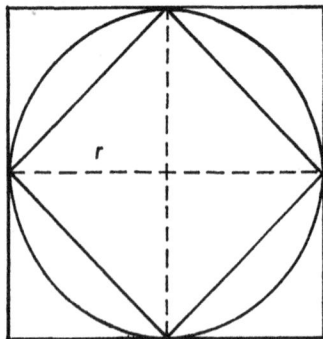

Our last figure on areas shows a circle contained in a square and containing a square, and it will help you to follow what is written above.

Table of units and applications

19. The Table of units in the Metric System is as follows:

Length		Area		Volume	
m.		sq. m.		cu. m.	
$\frac{1}{1000}$	1 millimetre (mm.)	$\frac{1}{10^6}$	1 sq. mm. (mm.²)	$\frac{1}{10^9}$	1 cu. mm. (mm.³)
$\frac{1}{100}$	1 centimetre (cm.)	$\frac{1}{10^4}$	1 sq. cm. (cm.²)	$\frac{1}{10^6}$	1 cu. cm. (cm.³)
$\frac{1}{10}$	1 decimetre (dm.)	$\frac{1}{10^2}$	1 sq. dm. (dm.²)	$\frac{1}{10^3}$	1 cu. dm. (dm.³)
1	1 metre (m.)	1	1 sq. m. (m.²)	1	1 cu. m. (m.³)
10	1 dekametre (dam.)	10^2	1 sq. dam. (dam.²)	10^3	1 cu. dam. (dam.³)
100	1 hectometre (hm.)	10^4	1 sq. hm. (hm.²)	10^6	1 cu. hm. (hm.³)
1000	1 kilometre (km.)	10^6	1 sq. km. (km.²)	10^9	1 cu km. (km.³)

It allows us to find the value of a unit in terms of any other in the same column.

Write down any unit and find all its expressions in terms of the others.

For example:

1 mm.$=\frac{1}{10}$ cm.$=\frac{1}{100}$ dm.$=\frac{1}{1,000}$ m.$=\frac{1}{10,000}$ dam.$=\frac{1}{100,000}$ hm.$=\frac{1}{1,000,000}$ km.

The circumference of the Earth is approximately 40,000 km. Find what it is in any of the other units of length.

The average distance of the Earth to the Moon is approximately 60 times the Earth's radius, which is about 6,350 km.

Find the distance of the Earth to the Moon in the various units.

The average distance of the Earth to the Sun is 150 million km. Give that distance in the various units of length.

II

CAPACITY AND VOLUME OF ROUND BODIES

PART II

CAPACITY AND VOLUME OF ROUND BODIES

Capacity

1. *Capacity* is another name for volume but it is used in connection with vessels intended to contain something, whereas volume relates to any portion of three-dimensional space occupied by a body.

A box is a prism and its volume can be calculated by the method explained in Part I, but we are more likely to be thinking of what it can contain than of the space it occupies. A tank, a milk-bottle, and a can have volume, but, here again, we are concerned with their contents and speak of their capacity though, for example, in the case of a tank we should refer to the volume of water it holds. So, capacity and volume are different names for the same thing, and units with special names are used to express capacity.

The litre, its multiples and submultiples

2. In the Metric System, the main unit of capacity is the *litre* which corresponds to the dm.^{3}. Its multiples are the

dekalitre (dal.), hectolitre (hl.), kilolitre (kl.).

Its submultiples are the

decilitre (dcl.), centilitre (cl.), millilitre (ml.), each being related to the one before as in all metric units.

For wholesale liquid measurement, the hectolitre is the commonest unit. For wine, or other liquids sold retail, dcl. are used, while ml. are used by chemists.

Find, using fractions, decimals, or powers of 10, the relationships between each of these units to all others.

This Table sums up all the relationships to the litre:

1 ml.	or	$\frac{1}{1,000}$ l.	corresponds to	1 cm.^{3}
1 cl.	or	$\frac{1}{100}$ l.	,,	$\frac{1}{100}$ dm.^{3}

1 dcl.	or	$\frac{1}{10}$ l.	corresponds to	$\frac{1}{10}$ dm.³
1 l.			,,	1 dm.³
1 dal.	or	10 l.	,,	10 dm.³
1 hl.	or	100 l.	,,	100 dm.³
1 kl.	or	1,000 l.	,,	1 m.³

Round bodies

3. No. 2 covers all one need know of the metric system in respect of capacity. In the remainder of this chapter, we shall consider the cylinder, the cone, and the sphere, which are common round bodies and, assuming they are hollow, we shall find their capacity. It is obvious that we shall at the same time be finding the volume of these important figures.

A cylinder can be made by rolling a sheet of paper as shown in the figure. Two sides of the rectangle become the two circumferences forming the edges of the cylinder. If we stick with glue the edges of the paper that meet and cut two cardboard circles of the appropriate size to fit each circumference, we form a hollow, closed cylinder.

Each straight line drawn on the cylinder, perpendicular to the circles, is called a *generator*. The two circles are called *bases*, and the line that joins the centres of the bases is called the *axis* of the cylinder.

We can generate a cylinder in another way:

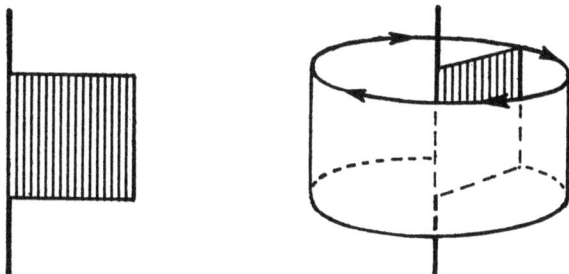

21

Make a rectangle of cardboard or stiff paper and, using one of its sides as axis, let it make a complete revolution. It is clear from the figure that the edge parallel to the axis actually generates the surface of a cylinder; hence the name 'generator'. The length of the generator is called the *height* of the cylinder.

A cone can also be made from a piece of paper. This time we start with a circle. Make a cut along two radii and roll it so that the two edges of the cut meet as shown in the figure. The circle of the base of the cone is made of the portion of the circumference of the given circle that is left after the cut.

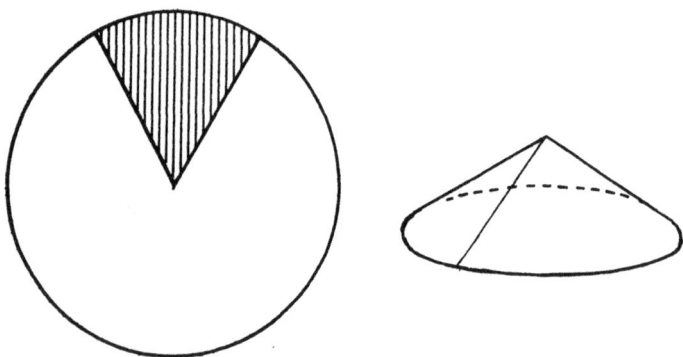

The centre of the given circle becomes the *apex* (or *vertex*) of the cone. The line that joins the apex to the centre of the base is called the axis of the cone. Each straight line joining the apex to a point on the circumference of the base is called a generator, and we can justify that name by making a right-angled triangle revolve around one of the sides of the right angle as in this figure:

Because the cylinder and the cone can be made with a flat piece of paper they belong to the same family of bodies, and we can see that the further the apex of a cone is from its base the more like a cylinder the cone becomes, whereas, the closer the apex is to the base the more the cone approximates to a circle. Indeed, a circle is a cone whose apex is on its base. Calling the distance of the apex to the centre of the base the *height* of the cone, we can say that a circle is a cone whose height is equal to zero.

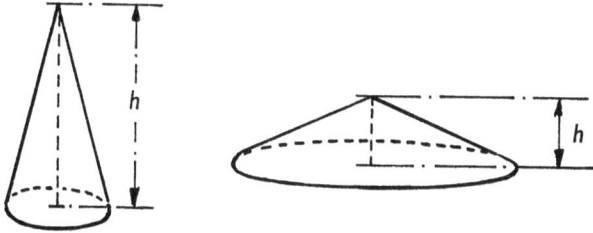

4. A sphere cannot be made out of a piece of paper, but can be generated by a semi-circle making a full revolution round its diameter, as in the figure below. The centre of the diameter of the semi-circle is the centre of the sphere; the radius of the sphere is the radius of the generating semi-circle. While a cylinder and a cone have one axis, every straight line through the centre of a sphere can be considered as its axis.

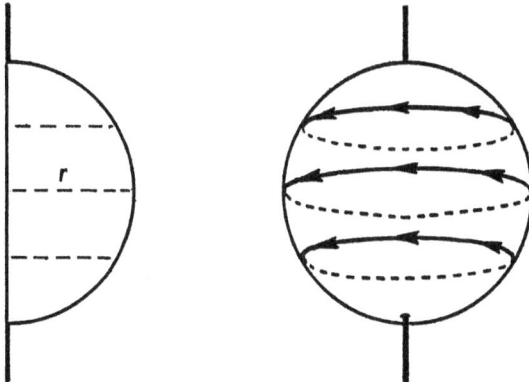

If an axis has been chosen, the circles perpendicular to it are called *parallels* and the one containing the centre of the sphere is the *equator* corresponding to that axis. The points of the sphere where the axis pierces the sphere are called *poles*. All circles of the sphere having the axis as diameter are called the *meridians*.

5. The capacity of a cylinder whose base is fixed will clearly vary with its height. If we fill a cylinder with water, we shall need twice the amount of water to fill it to a height $2h$ than to a height h.

The capacity of a cylinder whose height is fixed will vary with the width of its base (i.e., with the radius of the base). We saw in Part I (No. 18), that the area of a circle is equal to πr^2, where r is the radius and π a decimal number starting with 3·141. So we shall write that the capacity of a cylinder in c.c. is equal to the product of its height in cm. by the area of its base in square cm. or

$$V \text{ cylinder} = \pi r^2 h$$

expressed in the appropriate units. This will be proved to you one day.

We write V for volume or capacity because this is customary.

Find the height of a cylinder which must contain just one litre, if the diameter of the base is equal to 10 cm.

Find the radius of the base of a cylinder containing just one litre if its height is 20 cm.

Find the volume of a cylindrical tank whose height is equal to the diameter of its base and this equals 1 m.

6. We cannot as easily see with a cone as we could with a cylinder that its capacity is proportional to its height. If we

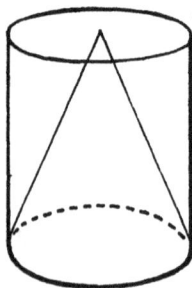

24

were filling a conical vessel with water, the amount of water needed to fill it to half its height would be considerably less than we should need to add to fill it to the brim. If we make a cylinder having the same base and the same height as our cone, we can place the cone in the cylinder and picture the difference as in the figure. By pouring water into the cylinder to fill that difference, we find that twice the amount is needed than is required to fill the cone. Thus,

$$V \text{ cone} + 2 \, V \text{ cone} = V \text{ cylinder} = \pi r^2 \times h, \text{ or,}$$

$$V \text{ cone} = \tfrac{1}{3}\pi r^2 h$$

using appropriate units for the two lengths r and h and the corresponding units for the volume. Thus, V will be in c.c. if r and h are in cm.

Answer the same questions as in No. 5 replacing the word cylinder by cone.

7. The capacity of a sphere is much more difficult to calculate by elementary methods. It is obvious that, since the only factor that determines a sphere is its radius, this alone will appear in the formula for the volume. It is also clear that since the volume of a prism is obtained by multiplying its three dimensions, the volume of a sphere contained in a cube with sides $2r$ (see figure) will have a volume smaller than $(2r)^3$ or $8r^3$. V sphere $< 8r^3$.

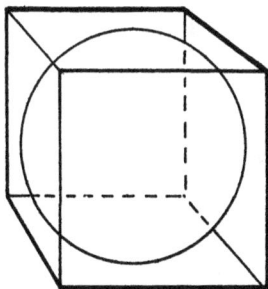

Later on, in your studies, you will find that the **exact** volume of a sphere in terms of its radius is given by

$$V \text{ sphere} = \tfrac{4}{3}\pi r^3$$

V is in c.c. if r is in cm.

25

Find the volume of the Earth, assumed to be a sphere, if you know that its radius equals 6,370 km.

How many spheres with radius equal to a quarter of the side of a cube can you place inside this cube?

Do you need to know the formula of the volume of a sphere to find the answer to that question?

Solids of revolution

8. Since we saw that the cylinder, the cone, and the sphere can be obtained by a complete revolution of a rectangle, a right-angled triangle and a semi-circle respectively, they are called *solids* (or *volumes*) *of revolution*. We may want to form more of these bodies, and here are some examples for you to study.

Try to see all the volumes of revolution you can obtain by making a triangle like the one in the figure below revolve around each of its sides or around a given axis, as shown:

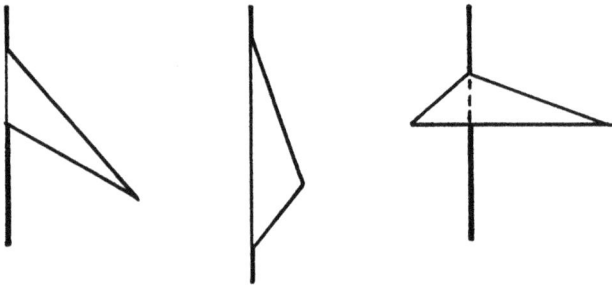

What is the body of revolution you can obtain through a complete revolution of

(a) a trapezium, round the axes indicated below?

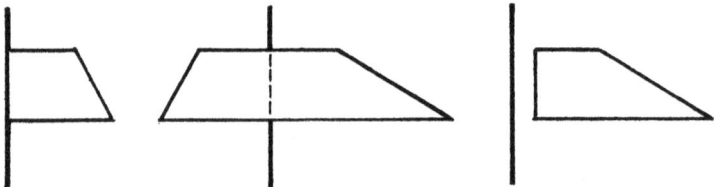

26

(b) a parallelogram?

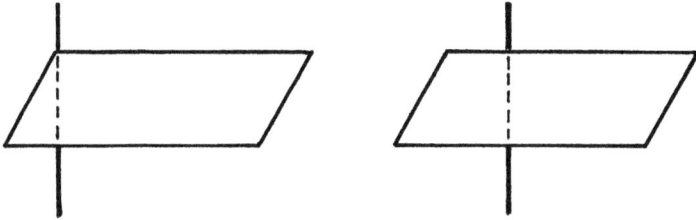

In all these cases, see if you can find the formulae for the volumes by the addition or subtraction of the volumes of the cones you can see in the figure.

If you cut a cone along a section parallel to the base or perpendicular to its axis, you obtain a circle. That part of the cone between the base and the section is called a *frustum*. Can you see some in the figures you have obtained?

If a circle is made to go round an axis in its plane, which does not touch the circle, a volume like the inflated inner tube of a bicycle tyre is obtained. It is called a *torus* and you can see it being generated in the following figure:

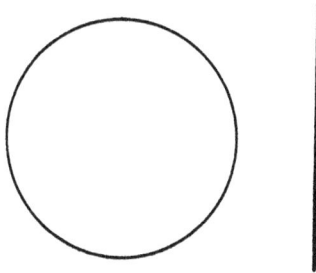

If the axis comes closer and closer to the circle, can you imagine what the resulting toruses will look like?

27

c

III

WEIGHT, MASS AND DENSITY

Part III

WEIGHT, MASS AND DENSITY

Weights and balances

1. We have all seen a balance, but we may not all know how it functions.

The first balance that was invented was just a rod at the centre of which a nail or a hook was placed and, when it was suspended from that hook, the two parts of the rod were horizontal. If the hook is not exactly at the centre, we cannot make the rod remain horizontal without pushing one side up or the other side down.

Archimedes, who studied the lever and gave the expression of its law, noticed that if he placed a flat piece of wood, for example a small plank, on a sharp edge and the parts were of unequal length, then the two parts could be made to balance if he placed a stone of suitable weight on the shortest side. He expressed this in the formula of the lever

$$P \times a = Q \times b$$

where a and b are the lengths of the two arms and P and Q the weights. If a becomes smaller, then P must be greater. If P becomes smaller, then a must be greater.

The Romans used a balance that had one long arm and one very short one, pivoting on a sharp edge. On a tray hanging from the small arm they placed stones and, at the end of the long arm, they hung the objects (mainly food) they wanted to weigh.

In the 17th century, in France, Roberval invented the scales with two equal and parallel arms joined at their extremities by two equal and vertical bars so as to form a hinged parallelogram. This principle is still used in scales for weighing letters. It ensures that the trays remain level.

Today, we often use complicated spring balances with only one tray, the weight being shown on a dial. But, here again, Archimedes' lever formula is the one that governs the relation between the weight and the movement of the marking needle.

If the balance is one with arms that are equal, we can say that only when the weights are equal will they balance, and if we have made a choice, once and for ever, of the objects we propose to use as units, then we can use the scales to tell what, in terms of these objects, any given object weighs.

Weighing

2. The problem of weighing has two aspects:

i) How can we find whether a body weighs twice as much as another?

ii) What units must we have to weigh as many bodies as possible with them?

It is clear that if we solve problem (i), we can find all the multiples of the weight of the given body, because, if a is the weight of the body we start with (and which we adopt as our unit), by balancing it exactly with another object, we find a again. Putting the two together, we obtain $2a$ which, balanced by yet another object, gives us $2a$ again. As we had a to begin with, we now can make a, a, $a+a$, $a+2a$. Using the balance again, we can find a, $2a$, $3a$ and, by putting these with those we had, we can form $4a$, $5a$, $6a$, etc. The only difficulty is that this would take too much time and we should not easily find objects that would exactly balance with a or $2a$.

Finding a fraction of a weight

3. Theoretically, as we have seen, we can find the multiples of any weight. Can we, equally, find its submultiples using the scales?

If the body we choose can easily be torn apart (say a lump of clay), we can cut it into two roughly equal parts, and put them on the scales. Generally, they will not balance exactly, and we shall know which is the heavier. We could then tear a

31

piece from the heavy part and add it to the other, but this may be insufficient to tilt the balance the other way. If so, we can again add some more clay taken from the heavier side, until equilibrium is reached or until the scales begin to tilt the other way. This method is lengthy and clumsy, but it illustrates the principle we are considering.

Another method is to use containers having volumes that are related to one another in appropriate fractions which we fill with water. If the containers chosen are of light material which do not take up more than a small percentage of the total weight when they are filled, we can use this method for obtaining whatever fraction of a given unit we desire. If, keeping the bottom of the vessel constant, we halve the height, then we also halve the weight; if we take a third of the height, we get a third of the weight, and so on.

4. Let us mark on a glass the levels reached by water when we pour into it 1 c.c., 2 c.c., 3 c.c. . . . of water, and let us put the empty glass on one pan of a balance and pour iron filings on the other to balance it. Then, each time we pour 1 c.c. of water into the glass, we can restore the equilibrium with more iron filings kept separate from those we put first. This gives us amounts of iron filings equal in weight to

1 c.c., 2 c.c., 3 c.c. . . .

of water. Placed in very light bags, they can now serve to balance other bodies instead of the water.

Now, it happens that the first unit of weight which was chosen in the metric system was the weight that exactly balanced with 1 dm.^{3} of water, and it was called the *kilogram*. From this relationship we can find all the multiples and submultiples of the kilogram we need by comparison with volumes. Since 1 dm.^{3}=1000 cm.^{3}, and $\frac{1}{1,000}$ kilogram is called a gram, we find that the iron filings which we placed in bags weigh 1 gm., 2 gm., 3 gm., etc.

By choosing a metal that can easily be moulded we can make solid metal weights that balance with one, two, three or more such bags of filings, and they can be marked with their weights in grams or other metric units of weight. These can easily be lifted, moved, and stored. We are familiar with such weights, which may be made of cast iron, brass, or some other copper alloy.

Units

In principle, we can make any weight we want but, among the submultiples and multiples of the kilogram (kg.), the following have been given special names in the metric system:

Name	Value in kg.	Value in gm.	Abbreviation
The milligram	$\frac{1}{1,000,000}$ kg.	$\frac{1}{1,000}$ gm.	mg.
The centigram	$\frac{1}{100,000}$ kg.	$\frac{1}{100}$ gm.	cg.
The decigram	$\frac{1}{10,000}$ kg.	$\frac{1}{10}$ gm.	dg.
The gram	$\frac{1}{1,000}$ kg.	1 gm.	gm.
The dekagram	$\frac{1}{100}$ kg.	10 gm.	Dg.
The hectogram	$\frac{1}{10}$ kg.	100 gm.	hg.
The kilogram	1 kg.	1,000 gm.	kg.
The myriagram	10 kg.	10,000 gm.	
The quintal	100 kg.	100,000 gm.	
The ton	1,000 kg.	1,000,000 gm.	

It is easy to translate any one of these units into any other. Express, as you did for lengths, areas, and volumes, as many as you can of these units in terms of the others, using fractions, decimals or powers of 10. Thus,

1 ton $= 10^3$ kg. $= 10^6$ gm. $= 10^9$ mg.; or 1 ton $= 10$ quintals $= 10,000$ hg.

1 Dg. $= 10$ gm. $= 0.01$ kg. $= 0.00001$ ton.

In business, tons, quintals, and kilos are the most common units, and the other weights are expressed as decimal fractions of these.

The smaller units, and decimals thereof, are used by chemists.

In shops, food and sweets are commonly sold by the kilo, $\frac{1}{2}$ kilo and hecto. The metric pound, or libra, is equal to 500 gm.

Making a box of weights

5. The second problem referred to in No. 2 is the one we have to solve if we want to make a box of weights containing the minimum number of weights capable of forming all the weights from 1 mg. to 2 kg.

This is a problem of arithmetic we can easily solve, perhaps using the rods.

We know that $1+2+3+4=10$, and that with 1, 2, 3, 4 we can form 1, 2, 3, 4, $1+4$, $2+3$, $2+4$, $3+4$, $1+3+4$, $2+3+4$, and $1+2+3+4$ i.e., all the numbers up to 10. But if we took 1, 2, 2, 5, what would the difference in arrangement be? We get:

1, 2, $1+2$, $2+2$, 5, $1+5$, $2+5$, $1+2+5$, $2+2+5$, $1+2+2+5$.

We again find all the numbers up to 10.

There may, however, be advantages in using one arrangement of weights as compared with the other. We can study this by seeing how many weights we need to weigh various bodies, first with one set of weights and then with the other. So let us make up two sets of weights based on the two systems, as follows:

First Box		Second Box	
mg.	1, 2, 3, 4	mg.	1, 2, 2, 5
cg.	1, 2, 3, 4	cg.	1, 2, 2, 5
dg.	1, 2, 3, 4	dg.	1, 2, 2, 5
gm.	1, 2, 3, 4	gm.	1, 2, 2, 5
Dg.	1, 2, 3, 4	Dg.	1, 2, 2, 5
hg.	1, 2, 3, 4	hg.	1, 2, 2, 5
kg.	1	kg.	1

In the sets of weights used in a laboratory, the smallest are made of aluminium rectangles which must be kept separately and picked up with tweezers. The bigger ones are placed in cylindrical holes in a block of wood.

Before we consider an actual weighing, let us answer the question: Do we need all these weights to weigh any body between 1 mg. and 2 kg.? The answer is yes, because in each decade we need the four to be sure we can compose any number.

If the chosen body weighs 747 gm., we should take the weights as follows:

	From First Box		From Second Box
For 700 gm. take	4+3 hg.	or	5+2 hg.
40 gm. take	4 Dg.	or	2+2 Dg.
7 gm. take	4+3 gm.	or	5+2 gm.

Here we see that there is a slight advantage with the first box because only 5 weights are needed as against 6. But this is not always so, because to weigh 555 gm. we should need 6 weights from the first box and only 3 from the second. However, for 444 the respective number of weights would be 3 as against 6, and for 333 it is again 3 as against 6. So there is an advantage in the first, but it is the second that has been chosen. The weights (ignoring the ones that must be picked up with tweezers) are arranged in the box as below:

1 gm.	2 gm.	2 gm.	5 gm.	10 gm.	20 gm.
20 gm.	50 gm.	100 gm.	200 gm.	200 gm.	500 gm.

A kilogram weight is also added.

The reason for this choice is that only three moulds are needed, instead of four, to make each group.

Find the weights you would need from this box to weigh 813 gm., 207 gm., 1472 gm., 179 gm., 53 gm., 999 gm., and 1547 gm.

Mass and weight

6. The word 'weight' we have been using is in fact not the correct one when we use scales. It would be correct if we were using the *spring balance* which is made of a spring that is pulled down by the actual weight of a body, this being defined by the attraction of the Earth on the body.

It has been discovered that the weight of a body is not constant but depends on where, upon the planet, the spring balance is used; the same body weighs more at the poles than at the equator, and its weight varies continuously if we move on a meridian, whereas it remains constant when we move on a parallel.

By the use of the pendulum, the time of whose oscillations depends upon the attraction of the Earth, it is possible to find how gravitation varies with the position on the planet and, knowing this, we can compare weight (variable with the position) and *mass* (which is constant) for every body.

The units of mass are the same as the units of weight, but it is usual to add, after kilogram, the words *mass* or *weight* to express which is the magnitude considered.

7. If we subdivide a body into several others, the mass of the original one is equal to the sum of the masses of the parts. If a body is not submitted to alterations that remove parts of it, its mass never changes.

Note that this property of the mass is not shared by the colour or the temperature of a body, which may well change without the body losing any of its parts.

Density

8. When we consider only a unit of volume of a substance, its mass is called the *density* or *specific mass* of that substance. In that way, we can compare bodies.

While the question: 'which weighs more, a kg. of lead or a kg. of feathers?' is a joke and a catch, the vague idea that feathers are 'lighter' than lead becomes precise when we compare the mass (or the weight) of the same volume of two substances.

In the following table, we give the density in grams per c.c. at 20°C of a few of the most common metals and a few other substances:

Aluminium	2·70	Silver	10·50
Copper	8·94	Tin	5·75 to 7·31
Gold	19·32	Zinc	7·14

36

Iron	7·87	Carbon	1·88 to 3·51
Lead	11·3	Phosphorus	1·82 to 2·70
Mercury	13·55	Sulphur	1·92 to 2·07
Nickel	8·90		

When there is more than one number, it is because there are several forms for that substance. Carbon, for example, is found as graphite and in the form of diamonds and has other forms as well, each of which has a different density.

The numbers in the Table tell us that one cm.^{3} of each substance weighs so many grams.

9. The method of finding the density of a substance is based on two experiments, one giving the mass of a body, and the other giving its volume. If the body has a prismatic shape, we could measure the appropriate dimensions and obtain its volume by multiplication. But if it has a very complicated geometrical appearance, we must find a way round. The method found was to immerse the body in a liquid (generally water) and, as it displaces a quantity of that liquid equal to its volume, all that is needed is to find the volume of the liquid so displaced.

In the case of pure water at 4°C, its density is *one* because 1000 c.c. weigh 1000 gm. So, using water is the most economical way of obtaining the volume of bodies of any shape in a short time.

Let us take a piece of any solid substance and weigh it. But, instead of putting it in one of the trays of the scales, we hang it underneath. This does not make any difference to its weight.

When equilibrium is established, we take a glass *full* of water on a saucer, and carefully introduce the solid into the liquid. The water it displaces will flow over into the saucer, and, if we weigh it, we shall find the number of grams equal to the number of c.c. collected. This will give us the volume of the substance immersed.

If we now divide the weight of the substance by the weight of the water displaced, we find the required density. All the numbers given above (in No. 8) were found by that method.

Applications

10. Tables of density are useful for finding the weights of given volumes made of those substances. A simple multiplication will give us the answer. Reference to them should be made in all practical problems.

Here is an example of a problem. In an alloy, made of two metals, it is known that copper is one of them. Can we find which is the other if we are told the density of the alloy and the proportion of the two metals in it?

Let us say the unknown metal is in the proportion of 25% of the alloy and that the density of the alloy is 7·38.

In each c.c. of the alloy, there are $\frac{75}{100} \times 8·94$ gm. of copper or 6·705 gm. Thus, there are 7·380—6·705, i.e. 0·675 gm. of the other metal. But that represents $\frac{25}{100}$ of 1 c.c. Hence, the density of the other metal is equal to 4 times that number or $4 \times 0·675 = 2·70$. It is that of aluminium, so the other metal is aluminium.

Chemical methods give the answer if one does not know exactly the composition of an alloy; but the method explained above is actually used in determining whether metals such as gold contain more than the amount of alloy permitted by law for any particular purpose.

In Part VI of this book you will find problems on mixtures and alloys that require a knowledge of this topic.

11. One more word about density. Bodies are made of atoms and molecules, and their mass is made up of these whereas their volume varies with pressure and temperature. Thus, the density of a substance is not a constant, except when the temperature and the pressure are maintained unaltered.

On the surface of the Earth we are accustomed to certain conditions that do not prevail everywhere in the Universe. In some parts of the atmosphere, the density is almost zero; at the centre of the Earth, the density reaches three times the average for the whole planet. In some stars, it is several thousand times the highest values known on Earth. This results from conditions of temperature and pressure that reduce the space between the nuclei of the atoms. It is therefore important to remember that not only do bodies have different densities but these densities are not always the same for the same substances; they change with the prevailing conditions.

To sum up, we have learned in this chapter that the instruments that measure masses are scale-balances, whereas weights are found by using spring-balances. The difference between weight and mass is to be attributed to the shape of the Earth: mass relates to the substance itself whereas weight results from the attraction exercised by the Earth upon it. The units are the same for both, but they must be qualified as units of weight or units of mass.

Density, or specific mass, is the mass of the unit of volume. It is the quotient of the mass divided by the volume, measured in the appropriate units. The density of bodies depends upon conditions of pressure and temperature.

IV

MEASURES, THEIR FORMALISATION

PART IV

MEASURES, THEIR FORMALISATION

Dimensional formulae

1. In the previous sections, we have met some of the most usual quantities of everyday life and have found that some of them are related to each other.

In this part of the book, we are going to study a wider theme: that of measures in general and of their formalisation.

To begin with, we shall consider the various conditions of measurement which we found were basic to all of them in the first three Parts. Then we shall consider other measurements we have not studied so far, and finally, we shall consider what are called the *dimensional formulae* for measures and physical magnitudes.

Measuring a length

2. When we are given a length and asked to find its value, we begin by trying to make up that length by placing a number of particular units end to end. If it happens that the length coincides with this series of units, we count the units and say the length L is equal to so many times the unit u

$$L = n \times u (n \text{ being a whole number}). \quad \ldots (1)$$

But, on the whole, it is most unlikely that the lengths will coincide in this way, and the length L will be bigger than $n \times u$ and smaller than $n \times u + u$ or $(n+1) \times u$ which we write

$$nu < L < (n+1)\ u. \quad \ldots (2)$$

Sometimes this information may be sufficient as, for instance, if we want to know how many tiles of length u are needed to make a frieze in a room or a building.

But if we are asked to find the value of L by measurement, we must use a unit smaller than u. By subdividing u into ten equal parts, we can take as a new unit $u' = \frac{u}{10}$. For example, we can take the mm. instead of the cm. Then, we put the new units end to end and see whether, this time, an exact number of u' is equal to L. This being very exceptional, we find again that L falls between $n'u'$ and $(n'+1)\,u'$ and we can write

$$n'u' < L < (n'+1)\,u'. \qquad \ldots (3)$$

You already know that if $u' = \frac{u}{10}$, n' must be bigger than $10n$, but it is worth testing this again to make sure you understand thoroughly why this is so.

Draw any length L. Measure it in cm. to obtain n. Then convert n into terms of $\frac{u}{10}$, which will give you $10n$ millimetres. Finally, measure L again in mm. to obtain n' and compare n' with $10n$.

Do the same for $(n+1)\,u$ and compare $10\,(n+1)$ with $n'+1$. Which is smaller? Why?

Usually, when measuring in cm. and mm. you use a *ruler* upon which the cm. and mm. are set end to end, and this saves the time that would be occupied by putting them end to end ourselves in the way we do when using white rods for measuring. But if the length to be measured is longer than your ruler, what would you do?

A given length L is usually not an exact number of cm. or mm., or even of a smaller fraction of either. So we have to accept for its *value* a number which is only an approximation of L. What we must do is to find the length to a given approximation as we did in Part I. We can then say:

$L =$ so many units to a fraction of that unit

meaning, as in (2) and (3), that if we add one more of these units the length obtained will be bigger than L.

3.   In using the units end to end, or in using the ruler, we have been very careful not to let them overlap or leave a gap in between. So these two conditions must be part of the process of measuring lengths.

In other words, we can say that *length* is a property of straight lines such that if s_1, s_2, . . . are a few such lines and $L\,(s_1)$, $L\,(s_2)$ . . . are the numbers respectively attached to

43

each as its length, L $(s_1)+L$ (s_2) will be, for example, the number attached to the line obtained by putting s_1 and s_2 end to end without overlapping or any gap between. If we call s_1+s_2 the operation of putting lines end to end, then we can write

$$L\ (s_1+s_2)=L\ (s_1)+L\ (s_2). \qquad \qquad ...\ (4)$$

This relation sums up what we do when we measure lengths of lines. Note that the number L is different when we use different measures, though the straight line s is always the same. For example, the length of this page is so many cm. or so many mm. (or so many inches), but it remains unaltered in size whatever unit we choose to measure it. s_1+s_2 is an operation on lines, while L $(s_1)+L$ (s_2) is an operation on number and numbers are added, not put end to end.

If we put several lines end to end without overlapping or gap, we can write, whatever their number, that

$$L(s_1+s_2+\ ...\ +s_p)=L(s_1)+L(s_2)+\ ...\ +L(s_p) \qquad ...(5)$$

which only expresses what we did in measuring. On the left-hand side the sign $+$ means: putting lines end to end, whereas, on the right-hand, it means: adding the appropriate numbers.

4. Is it true that if you draw a line AB and measure it from A to B, you get the same result as when you measure it from B to A? Do it on any line AB you wish to draw.

The question can be expressed:

Is $L(AB)=L(BA)$? $\qquad \qquad ...\ (6)$

If B comes closer and closer to A, the line is reduced in length, and if B coincides with A, we can say that the line is reduced to a *point*. The length of a point, we shall consider to be 0. If, instead of AB, we write AA to indicate that B now coincides with A, we can write

$$L(AA)=0 \qquad \qquad ...\ (7)$$

So length has three properties, which we can read in the three relations expressed as in (7), (6) and (4). The first says that the length of a line reduced to a point is 0; the second, that the length of a line is independent of the end at which the measurement is started or finished; and the third, that if a line is made of two bits that do not overlap and have no gap between, the length of the whole is equal to the sum of the lengths of the parts.

5. What happens to (4) if you make two lines s_1 and s_2 overlap?

What happens if s_1 and s_2 are two separated lines?

Isometric measures

6. In everyday life, we do not measure the lengths of straight lines only. Dressmakers measure waists, which are rounded, and use tape measures to do so. Because the tape can follow the shape of what is being measured, it can report the value of the length of that body, but the final measurement is made by stretching the tape as if it were a ruler and reading off the number found as if it had measured a straight line.

So, the measuring of non-straight objects is brought back to that of straight objects by means of tapes.

If you draw a curve freehand, and want to find its length, what do you do? You first find a string or thread to place upon the curve from one end to the other, making sure it follows the line you have drawn as closely as possible, and then you stretch it into a straight line against a ruler. In fact, what you do defines the length of the curve. You have found a way round the difficulty created by units of measurement being straight by using a thread which can take all shapes, including the straight line. It is like the dictionary we rely on when translating from one language into another, but it relates lengths of curved lines to lengths of straight lines. What you did can be said to be a *transformation* of the curve into a straight line, changing everything except the length. This type of transformation, we shall call *isometric* (from the Greek *same measure*).

By using this idea of isometric transformations, we can find the length of any curve which has a length.

45

Are there curves which have no length?

Yes, these are the curves so long that we cannot find their length and can only say they are *infinite*. For example, you can imagine a *spiral* that goes on and on, indefinitely; its 'length' cannot be found, though you can find the length of many portions of it.

7.   Take, instead of thread or string, a piece of elastic (a rubber band) with two ends. What happens to its length if you pull the two ends apart? Is there a length that is really that of the piece of elastic?

From this we learn that, in order to make proper units for measurement, we must find a material that does not vary perceptibly in varying conditions.

Measuring an area

8.   We saw, in Part I, that areas of plane figures can be measured using squares, but we only considered rectangles and figures composed of rectangles or readily reducible to rectangles. Can we do this with all figures?

Draw a closed curve having any shape you wish, but start with one that is not too complicated, such as the one below.

Can you draw a square that is completely contained in it? Note the biggest which has this property. Can you draw one that contains it completely? Note the smallest which does so.

If we call the portion of the plane enclosed by the curve A, the square inside it q_1, and the one containing it Q_1, we can say:

$$q_1 < A < Q_1$$

Such a statement is, however, unlikely to be of much value to us and, the more complicated the figure, the less value it will have.

Now subdivide the side of either square into two parts and use the lengths so made to construct four smaller squares within the one you chose. Subdivide the sides of these into two parts and make more squares using the new lengths. Do this again to make even smaller squares. As you subdivide in this way you will find a set of squares q_2 such that q_2 is contained in A but is bigger than q_1. You will also find Q_2 such that A is contained in Q_2, which, however, is smaller than Q_1. If you chose the outside square for this subdivision, the lines formed should be extended to make a mesh of squares covering the figure. The figures below will show what we are trying to do.

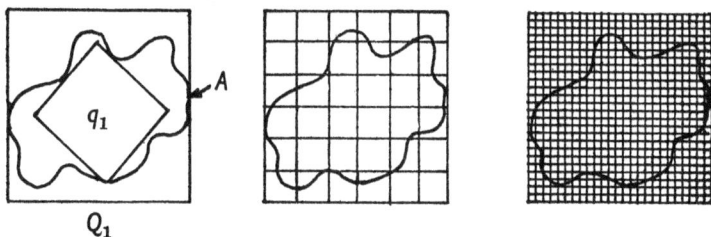

We have found that

$$q_1 < A < Q_1 \qquad \qquad \ldots (8)$$
$$q_2 < A < Q_2 \qquad \qquad \ldots (9)$$
$$\text{and} \quad q_1 < q_2 \text{ but } Q_2 < Q_1$$

This means that we can replace the squares (q_1, Q_1) by pairs of figures (q_2, Q_2) (q_3, Q_3) . . . composed of squares (thus having an area we can find simply by counting), and each time we make the squares in the sets smaller we can obtain new sets with an area 'nearer' to A.

This general method of approximation will give us 'better and better' values for A, but the true value of A cannot be exactly given by any pair.

The figures given above suggest to us that we could transfer the curved figure to graph paper to find its approximate area.

47

Transformation of a figure

9. If we start with a square F_1 equal to one unit, cut out a piece from one side and replace it upon another side (as shown), what is the area of the new figure we have made?

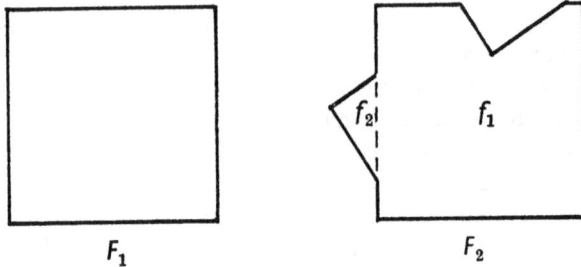

We see that F_2 (as we shall call this new figure) is made up of two pieces placed side by side with neither overlap nor gap between them. We shall call the piece we cut out f_2 and what was left after cutting this out f_1. So we can write, $F_2=f_1+f_2$. The sign $+$ means that the pieces have been joined without overlap or gap.

$A(f_1)$ and $A(f_2)$ being the areas of the two pieces, we can write

$$A(f_1+f_2)=A(f_1)+A(f_2) \qquad \ldots (10)$$

The sign $+$ on the right-hand side means that we are adding two numbers which indicate the value of two areas.

If from F_2 we cut another piece f_3 and replace it somewhere on one of the sides, we obtain a new figure F_3 whose area is still 1. F_3 is composed of f_3 and f_4 (which is what we shall call the piece that was left of F_2 after removing f_3). So we can write

$$A(F_3)=A(f_3)+A(f_4) \qquad \ldots (11)$$

This process of cutting off pieces and placing them elsewhere upon the figure will enable us to produce any type of figure we can think of and, for each of them, the area will always be equal to 1.

Equation (11) expresses for areas what equation (4) expressed for lengths. Look back and compare this for yourself.

A condition for measurement of areas is that the composing parts neither overlap nor leave a gap. Since we know how to find by direct means the areas of sets of squares or of rectangles, we can find, through them, as we did on p. 47, the value which we can call the *area* of a plane figure. This value, even if it cannot be found exactly by our method, is what we want because we can make use of this method to achieve whatever degree of accuracy we need.

What we have done in this section with the square F_1 gives us a method of obtaining as many figures as we want from a given one, each having the area of the first figure.

Now we have a much better idea of what areas are and of the supply of plane figures at our disposal.

Make as many successive figures as you wish from the square F_1 by cutting out pieces and sticking them in various places. After each step, copy the figure, so that you keep a record of your stock. Compare the figures you make with those of other pupils. Can you see how they obtained theirs? Have they observed, like you, the conditions of non-overlapping and absence of gaps?

10. One more experiment with areas can be tried here. Draw the three sides of a rectangle ABCD as in this figure:

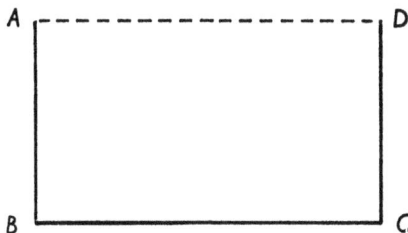

Then put a line across AD parallel to BC. You have formed a rectangle and you know it has an area. When you move the line AD towards BC, what happens to the area of the rectangle?

49

When AD coincides with BC, the rectangle is reduced to the line BC and the area is reduced to 0. This is a special case of plane figures, one with 0 area.

Of course, we can move AB towards CD at the same time as we are moving AD, so that the rectangle diminishes in all its dimensions and ends up by being the point C. Its area will again be zero.

Try to see what happens if, while AD tends towards BC as before, AB *moves away* from CD. Will the area diminish and tend to zero?

Measuring a volume

11. Volumes of three-dimensional figures can be treated as we treated lengths and areas to see what the conditions are for finding them.

Using the rods, we can form bodies that have straight sides, though not necessarily prismatic in shape, and it is possible to find their volume. In all cases, the rods are side by side and cannot overlap, nor are there gaps between them. Together, they form a continuous body.

Volume is a number attached to a three-dimensional body such that, if this body is made of two parts that neither overlap nor have gaps between, the three numbers V (the whole), v_1 (one part) and v_2 (the other part) satisfy the relation

$$V = v_1 + v_2$$

To make it look like equations (*4*) or (*11*), we can call F the whole body made of the parts f_1 and f_2:

$$V(f_1 + f_2) = V(f_1) + V(f_2) \qquad \ldots (12)$$

where the sign $+$ on the left says: put the two pieces together, while the sign $+$ on the right tells us to add the two numbers.

In a prism, we can imagine that the top face is moving and comes closer and closer to the bottom face, so that the volume diminishes all the time. When the two faces meet each other, there is no longer a three-dimensional figure and the body is said to have a volume equal to zero.

50

Measure; its properties

12. So far, we have found that length, area, and volume are properties of figures which satisfy the relationship:

$$m(a+b)=m(a)+m(b) \qquad \ldots (13)$$

which, on the right, tells us to add two numbers, while, on the left, it tells us to put, without overlap or gap, the two magnitudes together in the following ways:

(a) end to end, if they are lines,

(b) side by side, if they are plane figures,

(c) side by side, if they are three-dimensional figures.

This property of measurement is to be associated with the following statement

$$m(a+b)=m(b+a) \qquad \ldots (14)$$

which indicates that the order of arrangement does not affect the result, and with

$$m(O)=O \qquad \ldots (15)$$

which indicates that if a line is a point, or if a plane figure is a line or a point, or if a volume is a plane figure, a line, or a point, then the measurement is equal to zero.

Conditions (13), (14), (15) apply equally to capacity and to weight. Can you make sure of that?

They are the properties of all measurement and not only of those we have studied so far.

Other magnitudes

13. We shall now consider some of the other measurements one meets in everyday life.

Time is measured in various units—seconds, minutes, hours, days, weeks, years—and we know that, to measure time, we can use amounts of sand in hourglasses, water in clepsydras, or space on a sundial or clock-face, so that there is always something we can measure. But time is an experience that is different from space and weight, and we must consider it separately.

We can measure time as we measure length, that is to say we can write for an event that can be subdivided into two *successive* events (i.e. without overlapping or a gap between), that the time of each can be added to give the time of the whole event. If E is the whole event and e_1 and e_2 the other two, we can write

$$T(E) = T(e_1 + e_2) = T(e_1) + T(e_2) \qquad \dots (16)$$

For example, the time occupied by a meal can be made up of the time for eating and the time for drinking. So even an event like this can be subdivided into two or more non-overlapping, non-separated events.

We have to note that, with time, we *cannot* reverse the order of events, so that there is no clear meaning for

$$T(e_1 + e_2) = T(e_2 + e_1) \qquad \dots (17)$$

unless we mean that we are using the other formula: $T(e_1) + T(e_2) = T(e_2) + T(e_1)$ indicating that addition of numbers is commutative; but this we already know and it does not tell us anything about time.

14. *Temperature* is also a special experience but, here, with a thermometer, we can prove that temperature is not a magnitude satisfying the relations of measurement of No. 12. Indeed, if we mix water at 15°C with water at 14°C, we can obtain, according to the respective quantities mixed, any temperature between 14° and 15°C.

In order to make temperature a measure, we would have to introduce very complex ideas that are only met with much later in the study of physics.

15. What about *speeds* or *velocities*? Do they satisfy the conditions of No. 12?

First of all, speeds have direction and, if we consider only speeds in the same direction, we shall have to deal with positive and negative speeds. This for us is a new idea.

What do we mean by the addition of speeds? Can a body have two speeds at the same time? To see what this means, we have to consider one body moving with respect to another, which is itself moving with respect to a third. For example, a car on a road and a fly in the car. If the fly goes in the direction of the car, then its speed with respect to the road is the sum of the two speeds: its own with respect to the car, and the car's with respect to the road. So, speeds in certain circumstances satisfy condition (*13*) of No. 12. They then will satisfy condition (*14*), and if there is no movement at all, condition (*15*) is also satisfied.

But, in general, velocities, because they have direction, do not satisfy these conditions; velocities have measure but are not a measure themselves.

For instance, if two people go separately at speeds of 4 and 5 km. per hour, how fast do they go if they travel together? This is a nonsensical question as we know from everyday experience.

Scalar and vector quantities

16.   The magnitudes that satisfy conditions (*13*), (*14*), (*15*) are called *scalar*. The others, except temperature, are *directed* or *vector*.

Make a list of the scalar and directed quantities or magnitudes you can find and give the units in which they are usually measured.

For example, velocities, in a certain direction, are measured in km. per hour for cars, in km. per second for missiles. Find more examples and different units.

Dimensional formulae

17.   Measurements can be classified in another way, too.

You have noticed that lengths, areas, and volumes relate to one-dimensional, two-dimensional and three-dimensional figures. To measure the area of a rectangle, we use two dimensions: the length and the breadth. To measure the volume of a prism, we use three dimensions: the length, the depth, the width.

If L represents a one-dimensional *length*
 L^2 will represent an area of a two-dimensional figure.
 L^3 will represent a volume of a three-dimensional figure.

53

Mass cannot be reduced to length, nor can *Time*. (The phrase 'length of time' is, of course, only a manner of speech.)

Time is represented by T and Mass by M.

Formulae such as area$=L^2$, volume$=L^3$ are called *dimensional* formulae.

Let us find a few dimensional formulae for other magnitudes we meet in everyday life.

Velocity. As you know, speed in a certain direction expresses (for example) the number of km. covered in a certain time. We find it by dividing distance by time. But distance is a length.

Hence Velocity$= \dfrac{L}{T}$

This can be expressed as $L \times \dfrac{1}{T}$ and $\dfrac{1}{T}$ is often written as T^{-1}, so the dimensional formula of speed is

Velocity$=LT^{-1}$

Acceleration. On a car, or any moving body, an increase of speed is known as acceleration. It is measured by the amount the speed increases per second, or by dividing the speed by the time taken to reach it, assuming that we reach it steadily.

Acceleration$=\dfrac{\text{Velocity}}{\text{Time}}$ and the dimension formula is

Acceleration$=\dfrac{L\,T^{-1}}{T}=L\,T^{-2}$

To follow this you need to know that, just as $\frac{1}{T}$ is written as T^{-1}, so $\frac{1}{T^2}$ is written as T^{-2} and $\frac{1}{T^3}$ would be T^{-3} etc.

Force. An important formula you will need very often in your studies of science is Newton's, linking force, mass and acceleration: $F=M \times a$. Hence, the dimensional formula of Force is

Force$=$Mass $\times$ acceleration

$\qquad =M \times L\,T^{-2}=M\,L\,T^{-2}$

Pressure. This is the force exerted per square unit of the surface of bodies, or Pressure=Force/surface area.

The dimensional formula for pressure is

$$\text{Pressure} = \frac{M \, L \, T^{-2}}{L^2} = \frac{M \, T^{-2}}{L} \text{ or } M \, L^{-1} \, T^{-2}$$

Exercise

18. Given that
   a) Density=Mass per unit of volume
   b) Work=Force through a certain distance
   c) Power=Work per unit time
   d) Flow=amount of matter passing a surface per unit of time

find the dimensional formulae of

   Density
   Work
   Power
   Flow

In this book, we cannot introduce other magnitudes which contain the electrical and magnetic units since we cannot easily explain them. But all of them have dimensional formulae.

19. The use of these formulae is most important for the engineer and the scientist. We have included them here because they are really easy and because this book is about units and measurements. Later on, when you need these units you will know where to find them.

V

THE C.G.S., M.K.S. AND M.T.S. SYSTEMS OF UNITS

Part V

THE C.G.S., M.K.S. AND M.T.S. SYSTEMS OF UNITS

Historical notes

1. In this short chapter, we shall consider the various systems of units which have been proposed as the basis of the metric system. They are all linked together, but are preferred one to the other according to the field in which they are applied.

During the Middle Ages, the peoples of the world had difficulty in communicating with each other, and units, even in various parts of one country, were often very different and diverse. Even now, there is no uniformity in the choice of the units throughout the world, though scientists are steadily adopting the C.G.S. (i.e. centimetre, gram, second) system of which we shall say something soon. In Book III, we considered the British system of units which is a simplified version of even older systems.

During the French Revolution and Napoleon's reign (1789–1815), a radical alteration in the systems of units took place, which can be characterised by two words: simplicity and uniformity.

The decimal system, of which we have seen so many examples, was adopted for use in respect to all magnitudes. This is in great contrast with the previously prevailing systems and, particularly, with the present British system.

2. The French Academy had, during the early 18th century, undertaken two geodetic surveys to find the shape of the Earth. It was then believed that this could be simply determined, and it was already known that the Earth was not a sphere, but very nearly so. So, by a Bill on measurements enacted in 1791, the French Assembly, ordained that the unit

of length, or *metre* (from the Greek to *measure*), be linked to that natural magnitude, the circumference of the Earth. Thus, the metre was defined as the 40,000,000th part of the Earth's meridian. A piece of platinum was marked with that measurement and became the standard to which all other metre measures were to conform, and with which they could be compared to detect possible discrepancies.

It is now known that this definition of the metre is not accurate; it was decided, in 1919, to establish as the standard metre the distance between two marks on the bar of platinum, which was kept near Paris, and all reference to the meridian of the Earth was abandoned.

3.   By the same Bill, the French Assembly decreed that the unit of weight, or *kilogram*, should be that of a dm.^{3} of pure water at 4°C, this being the temperature at which water has maximum density. This links the kilogram to the metre, since the vessel has to be measured accurately. Later, weight was replaced by *mass*.

The third unit was the *second of mean solar time*, which is given by perfected clock systems maintained under special conditions in certain observatories. Nowadays, we make use of phenomena more precise than the clocks used in observatories to measure, with greater accuracy, the unit of time. For example, the vibrations of certain crystals are much more regular in their succession than the movements of pendulums or of the springs in clocks.

The system M.K.S. (metre, kilo, second) was the first to achieve uniformity in measurements, and it was soon adopted very widely.

Fundamental and derived units

4.   In that system, the most common units are the following:

Length: millimetre, centimetre, decimetre

metre

dekametre, hectometre, kilometre.

For *area* and *volume*, the square and cube of these measures are used.

59

Mass: gram, dekagram, hectogram,
kilogram,
myriagram, quintal, ton.

Time : second, minute, hour (not decimal).

L, M, T are the *fundamental units.*

The *derived units* are

Area ⎫
Volume ⎭ Derived from length as mentioned above.

Capacity: millilitre, centilitre, decilitre,
litre=1 dm.^{3}
dekalitre, hectolitre, kilolitre.

Density: kilogram per cubic metre or gram per cm.^{3}

Pressure: kilogram per square metre or kg. per cm.^{2}

Force: kilogram weight.

Work: kilogram metre.

Power: kilogram metre/second.

Heat: kilogram-calorie.

Most of these units are well adapted for trade and government, and were introduced mainly for that purpose. It
happened that some scientific units were also well represented
by these, but, in the case of others, it was too difficult to
effect conversions from one unit to another. Scientific
problems, which were discovered after the system was
established, created a demand for change, and a new arrangement of the M.K.S. system was proposed and widely accepted.
This is the C.G.S. (centimetre, gram, second) system, and
we shall discuss this in No. 5, but before we look at it more
closely, let us notice one example of agreement between the
M.K.S. system and a natural phenomenon.

In Europe, on some days at sea level, the mercury in the
barometer reaches 760 mm. in height. This is called the
normal pressure, and the weight of that column of mercury is
called an *atmosphere*. If we multiply the volume of 76 cm.^{3}
by the density of mercury in these circumstances, namely,
13·6, we find that the weight of that column is 1·033 kg. per

cm.^{2}, or very little above 1 kg. per cm.^{2}. Often we refer to an atmosphere as a kg. per cm.^{2}.

The C.G.S. system

5. The C.G.S. has, as its *fundamental units*, the centimetre, the gram and the second; the derived units that are of importance, but which are not easily deduced from the previous table, are

the *dyne*, which is the unit of force,

and the *erg*, which is the unit of work.

The dyne is the force which gives a mass of 1 gram. an acceleration of 1 cm. per second in one second.

The erg is the work done by one dyne acting through a distance of 1 cm. in the direction of the force.

The gram-calorie, or unit of heat needed to take the temperature of one gram of pure water from 14° to 15°C, is found to be equal to $4·185 \times 10^7$ ergs.

In the C.G.S. system, the name for the unit of pressure is the *barye*, and its value is one dyne per cm.^{2}. The megabarye, or million baryes, equals almost one kg. per cm.^{2} or an atmosphere.

Atomic measurements required the introduction of a suitable unit of length: the *ångström* (abbreviation *Å*), which is equal to 10^{-8} cm. or one hundred millionth of a cm. The dimensions of the atom are, for instance, of that order, whereas the nucleus has a radius one hundred thousand times smaller, or 10^{-13} cm.

The great importance of the C.G.S. system is found in its application to electrostatics, magnetism, and electromagnetism.* But we cannot study these in this book.

The M.T.S. system

6. More recently in the history of systems of units the M.T.S. (metre, ton, second) system was introduced. It is

---

* A most unexpected fact was discovered by Clark Maxwell in 1865 when comparing the electrostatic C.G.S. units with the electromagnetic C.G.S. ones. He found that the speed of light was the only magnitude in their ratios. From that followed a theory that made possible the discovery of radio waves and, therefore, of wireless and television.

easy to pass from the other two systems to this one: work out the tables of conversion.

The particular value of this new system is found mainly in technology where the units correspond better to the magnitudes considered.

For instance, the *therm* is the heat energy needed to increase the temperature of one ton of water by 1°C. So it is equal to a million gram-calories, or one megacalorie.

The *sthene* is the unit of force, or the force needed to give a mass of one ton an acceleration of 1 m. per second in one second. Since 1 ton$=10^6$ grams and 1 m.$=10^2$ cm., the sthene is equal to 10^8 dynes.

The M.T.S. system is still rarely used and only specialists need a good knowledge of it. Its inclusion here is for the sake of information. If you need it, you will know where to find it.

VI

PROPORTIONS AND MIXTURES

PART VI

PROPORTIONS AND MIXTURES

Sharing

1. Let us start with a length of 1 metre made of, say, 10 orange rods. If we replace the last rod with a light green and a black, we can write the following:

$$100 = 97 + 3 \text{ or } 100 = 93 + 7$$

Now imagine this expressed as (a) fractions, (b) decimals or (c) percentages:

(a) $\frac{3}{100} + \frac{97}{100} = 1$ or $\frac{7}{100} + \frac{93}{100} = 1$

(b) $0 \cdot 03 + 0 \cdot 97 = 1$ or $0 \cdot 07 + 0 \cdot 93 = 1$

(c) $3\% + 97\% = 100\%$ or $7\% + 93\% = 100\%$

We see that a single situation such as the one we started with can lead to a variety of expressions, notations, and word problems. If we add to that the properties of the operations which lead to inverses, we see that many more situations and many problems can be studied.

In this part of the text we shall explore these situations, leaving much unsaid. In each case you could find out more for yourself by reconsidering the situation and making your own alterations to it.

We can see that any given length can be partitioned into two or more smaller lengths and that it is possible to partition in the same way capacities or weights or any other additive magnitude (see Part IV).

This first remark is important, for what we can easily understand through using the rods will lead us to an understanding of other magnitudes we meet in our daily life.

2. For example, let us make with our rods, using only two in each case, lengths equal to the light green and the orange put end to end. We get black and dark green, or tan and yellow, or blue and pink or their reverse arrangements.

We can suggest many problems related to this situation of which the following are a few:

(a) Taking any multiple of 13 and assuming it to be a number of objects (fruit, sweets, acres of land, etc.) or units of money, share that number or that amount in the proportion represented by the ratio of the lengths of any of the two rods that form the length above. For example:

(i) Share £78 in the proportion of 6 to 7. The answer is £36 and £42, found by multiplying 6 and 7 respectively by 6, which arises from $78 \div 13$.

(ii) Share 6·5 acres in the proportion of 8 to 5 (consider $6\cdot5 \div 13 = 0\cdot5$).

(iii) Share 260 bags of coal in the proportion of 9 to 4.

Make up your own examples using any length and its partition into two complementary lengths. After you have made a few, reconsider what you did. Can you see that the following expressions show how you have been operating?

Let L be a length such that $L = a + b$

If we share $m \times L$ in the proportion of a to b, the answers are $(a \times m)$ and $(b \times m)$. Or written otherwise, they are:

$$\frac{m \times L \times a}{a+b} \quad \text{and} \quad \frac{m \times L \times b}{a+b}$$

These writings express what we have been doing in all cases

(b) Any number can be compared to any other, and we can always write: a is equal to the fraction $\dfrac{a}{b}$ of the

65

$$a = \frac{a}{b} \times b$$

magnitude b (of the same kind), or: $a = \frac{a}{b} \times b$

Similarly, 27 is $\frac{27}{13}$ of 13   11 is $\frac{11}{13}$ of 13

We thus see that it is always possible to repeat what we did in (a) above with actual multiples and submultiples of 13, in the case of any number m.

For example, we can share £11 between two people in the proportion of 6 to 7. The shares are £$\frac{66}{13}$ and £$\frac{77}{13}$.

Do you see that the operation applied in the two expressions is exactly the same, namely:

(1) $\dfrac{a \times m}{a+b}$ and $\dfrac{b \times m}{a+b}$ where m is any number

(2) $\dfrac{6 \times 11}{6+7}$ and $\dfrac{7 \times 11}{6+7}$

In the case of (1), m is the given thing to share, and a to b the given proportion; in the case of (2) £11 is the given amount to be shared in the proportion of 6 to 7.

Films and tapes

3.   For proportions which do not represent sharing, we have to consider how the given quantities relate to one another, and in problems of this type, we have to consider other factors, such as length of time, cost of materials, weight, etc.

We shall consider here films and tapes which we already know through having seen and heard them. This will give us some experience of problems about this type of proportions.

4.   You know that a film in a camera is affected by light, and that as a result we have photographs. We use films in ribbons on reels to make motion pictures. The eye sees a continuous picture if the films are run at a speed of at least 16 frames per second; if run at a lesser speed the effect is jerky.

(a) If a motion picture lasts for 90 minutes and there are 24 frames shown every second, how many frames have been taken in all? Silent 16 mm. pictures are shot and projected at the rate of 16 frames a second. What is the duration of the silent version of the same movie?

(b)　The films used in cinemas have frames measuring 35 mm. × 24 mm. but on 16 mm. films the frames are 16 mm. × 8 mm. How much more celluloid is needed to shoot the same story in the larger size than in the smaller, taking into account the two dimensions?

If a story shot on 16 mm. is projected in 7 minutes, what would be the duration of its projection had it been shot on 35 mm. film? What would be the lengths of the two films be?

(c)　On the celluloid of the films there is an emulsion which is sensitive to light and in some, to colour. Find the proportion of the area of film needed to produce the motion pictures in the two sizes given above. If 100 square cm. of film cost 1 penny, what is the price of unexposed film in both cases (i.e. 16 mm. and 35 mm.) for a 10 minute picture of the larger size?

(d)　We have a film shot on 35 mm., and we want to make a 16 mm. copy of it. We shall assume that we must project both at the rate of 24 frames a second. Find out the length of film needed for both, provided the 35 mm. one, when projected, lasts for 10 minutes.

Next, we have a 16 mm. film and we want to make a 35 mm. copy of it. Find the length of the film needed to do so provided the 16 mm. film when projected at the rate of 24 frames a second lasts for 10 minutes.

In terms of film material, how many times heavier or lighter will the new picture be, the thickness of the films being the same?

(e)　Apart from the cost of film, the postage charges will also be higher in the case of a 35 mm. film, and the reels and boxes used for it will be bigger and more expensive.

What do you think are the advantages of 35 mm. films which help offset the differences in cost mentioned above? If you know that a good picture depends on the emulsion and on the lenses of the camera used, you can see that a scene photographed on 35 mm. film will need 4 times as much emulsion as the 16 mm. film for the same object shot, and so images are sharper and richer in shades if taken on the larger size.

5. (a) Tapes for recording are used on apparatus having different speeds. The same length of tape at the ordinary speeds of 46·5 mm./sec., 93 mm./sec., 18·6 cm./sec., and 37·2 cm./sec. will last 16 minutes in the first case. Find its duration on the other speeds. What are the lengths of the tape used in each case?

Find the duration of 7440 cm. of tape at a speed of 18·6 cm./sec.

Find the duration of 1860 cm. of tape at a speed of 37·2 cm./sec.

Find the duration of 3720 cm. of tape at a speed of 46·5 cm./sec.

Find the length of a tape which at the speed of 9·3 cm./sec. would permit 16 minutes of recording.

Post Office

6. At the post office you can find that the cost of posting parcels, letters and the cost of telegrams and telephone calls are not always governed by the simple proportions we have met so far in this section. Look at the booklet which serves as a guide at your local post office. You will find in this booklet regarding inland postage:

(i) that it is usually proportional to weight with, perhaps, an exception for the minimum charge;

(ii) that even this proportion to weight is not strictly followed, for it takes into account some fractions: the relation is between such a weight and such and such a charge. (Prepare a table for these charge-weight relationships);

(iii) that the charges are not proportional to the distances of the places to which they are sent from your post office.

Overseas postage has a different rate but follows the same conventions. Is it true?

68

Air mail	air mail letters follow the convention stated above, but not air parcels, for their charges are proportional to weight and to distances.
Telegrams	are not uniform: (find the rule used for calculating the charges. What sort of proportion does it attempt to establish?)
Telephone	local, long distance, overseas calls are charged according to certain rules, (find them out and establish the proportion they suggest, if any).

Conversions of units

7. If we measure a length in cm. and again in mm., the two numbers obtained are in a simple relation: that of a proportion 1 to 10. It happens that the changes of units from the British to the Continental (or Metric) systems are governed by rules which are those of proportion or of proportion following a shift of the zero.

(a) For example:

$$1 \text{ inch} = 2 \cdot 540 \text{ cm.}$$

$$1 \text{ cm.} = 0 \cdot 3937 \text{ inches}$$

Since one thousandth of an inch is already quite small and only observable in precision work in ordinary life, we usually stop at the third decimal place: 1 cm. = 0·394 inches.

Express each of the following lengths in cm.

    1 foot   1 yard   1 furlong   1 mile

and each of the following in inches

    1 metre   1 dm.   1 km.

to the nearest unit.

(b) Do the same for areas and volumes and for capacity. Give a table of conversion worked out by yourself, one that allows conversion in both directions.

(c) Give a table of conversion of weights.

(d) Temperature is measured in degrees centigrade or Celsius in some countries and in Fahrenheit in others and these are written °C and °F. The same distance on the thermometer is divided into 100 parts for the centigrade graduation and into 180 for the Fahrenheit.

In addition to this, the 0°C is read 32°F. Find the formula for:

    (i)   converting centigrade temperature to the Fahrenheit.

    (ii) converting Fahrenheit temperature to the centigrade.

Check it on the two following correct correspondences:

$$100°C = 212°F$$
$$1000°C = 1832°F$$

Give the equivalent of the following temperatures:

37°C	30°C	15°C
106°F	97°F	0°F
120°F	3000°C	6000°F
1°C	1°F	−40°F

(e)   Since money exchange rates vary from time to time for unpredictable reasons, it is not possible to prepare a table valid for ever as it is for the other units. Still, if you wish you can make a table which gives the exchange values for units of money in various countries. Fill in the following *double entry* table, getting the rates from your newspaper (financial page).

U.S.$	£	F.F.	D.M.	S.F.	Fl.
U.S.$					
£					
F.F.					
D.M.					
S.F.					
Fl.					

F.F. French franc   D.M. Deutsch Mark   S.F. Swiss Franc
                         Fl. Florin (Holland)

On the basis of the table prepared by you, work out the following conversions of currencies:

$$\$75 = \pounds \qquad 132 \text{ Fl.} = \pounds \qquad 1{,}000{,}000 \text{ F.F.} = \pounds$$
$$\pounds 100 = \$ \qquad \pounds 32 = \text{S.F.} \qquad \pounds 1320 = \text{D.M.}$$
$$13 \text{ Fl.} = \text{D.M.} \qquad 50 \text{ D.M.} = \text{F.F.} \qquad 1000 \text{ S.F.} = \text{Fl.}$$

In these rates of exchange direct simple proportion is used. You can easily make a much wider table of conversion of currencies as there are over 100 of them used in the world.

Discounts

8. Here are a few examples of *discounts* which are allowed by traders to customers:

(i) 13 for the price of one dozen. What percentage is that?

(ii) 10% for a round-trip ticket by some air line companies, and 50% for children under 12. Students obtain similar discounts on certain flights. Calculate savings on certain journeys of your choice. Railways offer 20, 30 and 40% discounts to groups according to size and composition. For some special trains, on the other hand, there is an extra surcharge. Find examples and calculate costs.

Consider the cost of rail, sea or plane travel for some journeys, and find whether the charges follow a rule of proportionality. You will need to find distances first. In out-of-date travellers' guides you will obtain all the information you want. Get one from a local travel agent or air company.

There may be some reductions in charges: for group travel find which are offered to you in your region, and if they are not expressed in %, find the percentages in the various cases mentioned. Do they represent a proportionality?

9. Booksellers sell to the public and buy from the publishers. They sell books at an agreed price called the selling price and their profit is a percentage of that price, called the discount offered by the publisher to the booksellers. It varies from 10 to 50%. If authors get 10 to 20% of the same price, it follows that the publishers have from 30 to 80% of the selling price to cover all their expenses and to make a profit themselves.

Any book you consider has a price. This consists of the cost of paper, cost of composing the type, cost of plates and of illustrations, cost of printing, cost of advertising and publicity, office overheads, author's royalties, discount to booksellers, and when all that is covered, profit.

Consider this analysis of the price and find what happens to the profit of publishers, if the discounts remain the same. To get your answers consider each component of the price and see whether it involves the number of books printed or is an expenditure independent of that number. Double or treble the number of books sold and find out if the various items of costs will also double or treble?

So, in trade, the income of some people follows the rule of proportion, while for others it is a more complicated relationship which is not too difficult to calculate if we use additions, subtractions and proportions.

Compound unitary method

10.   In Book II we introduced the Unitary Method as an alternative to simple proportion. The *compound unitary method* is equivalent to *compound proportion* and is met in various problems in every-day life.

Here is an example of this type of question:

30 men take 4 days to dig a trench 200 m. long. How many men will be needed to dig a trench that is 500 m. long in only 2 days?

Two of the terms have changed here and we need to find a term that could be formed by direct proportion or unitary method if only one were changed. That is why it is called compound proportion or compound unitary method.

Using proportion we reason like this: In 4 days $30 \times \frac{5}{2}$ or 75 men would dig the 500 m. trench and in 2 days 150 will be needed. By compound unitary method we find what amount of the trench is dug by one man in one day. This is $\frac{200}{30 \times 4}$. The number of men needed to dig 500 m. in one day will be the number of times that this fraction goes into 500 or

$\dfrac{30 \times 4}{200} \times 500 = 300$; but in order to finish the work in 2 days, half of the number of men will be enough. We once again get the answer: 150 men.

We can write the question in two lines putting underneath one another the terms of the same kind:

(1)	200 m.	30 men	4 days
(2)	500 m.	? men	2 days

Compound unitary method will look as follows:

(1)	200 m.	30 men	4 days
(2)	1 m.	$\dfrac{30}{200}$ men	4 days
(3)	1 m.	$\dfrac{30 \times 4}{200}$ men	1 day

Do you see why this is the *unitary* line? Then:

(4)	500 m.	$\dfrac{30 \times 4 \times 500}{200}$ men	1 day
(5)	500 m.	$\dfrac{30 \times 4 \times 500}{200 \times 2}$ men	2 days

The answer is seen. A quick calculation gives 150 men.

11. In problems involving compound proportion any one of the terms of the second line above can be required to be found. It is always better to think of the problem as if we were actually doing what is required, so that we know which operation to perform, rather than attempting to solve the problem by guessing. For example:

(a) With 100 reams of paper, 150 texts of 320 pages each are made. How many reams are needed to make 120 texts of 280 pages each?

(b) How many reams are needed to make 200 texts of 350 pages each?

Work along the lines of compound proportion; the answers are:

(a) $$\frac{100 \times 120 \times 280}{150 \times 320} = 70$$ (b) $$\frac{100 \times 200 \times 350}{150 \times 320} = 145 \cdot 8$$

Can you solve these two problems by the compound unitary method, writing down the steps involved in each? Here are a few more examples to give an idea of how this type of problem can arise.

(i) To plough a field of 75 acres, 2 tractors need 2 days; in how many days will 5 tractors plough 120 acres?

(ii) To type the correspondence of 9 civil servants, 4 typists work 7 hours a day every day; how many typists are needed working 8 hours a day, to serve 7 civil servants of the same office?

(iii) If we furnish flats uniformly, £3,500 are needed to furnish three flats of 4 rooms each. What is the cost of furnishing a block of 20 flats of 3 rooms each (assuming that the cost of furnishing each room in the flats remains the same)?

12. Another type of problem that you can solve by the same method is the following:

Blending coffee of different qualities and prices leads to a mixture which will have a price intermediate between the cheapest and the most expensive.

Let us start with some bags of coffee of quality 1 at one price, and of quality 2 at another price. If we replace one bag of the quality 1 coffee by one of quality 2, the number of bags has not changed, but the total price of these bags has either increased or diminished. Let us say that it has increased. It will increase again if we replace one more bag of quality 1 by one of quality 2.

You can do that with your rods. Let us take 10 black rods and 10 blue rods, each rod representing the price of one bag.

the value of 10 bags of quality 1 will be 70 units

the value of 10 bags of quality 2 will be 90 units

Put the rods end to end to form a black and a blue train.

If we replace a black rod by a blue in one line and a blue by a black in the other we find the values (here the lengths):

74

88 for one and 72 for the other. If we continue we have:

86	,,	,,	74	,,	,,	
84	,,	,,	76	,,	,,	and so on until we get
70	,,	,,	90	,,	,,	

So between 70 and 90 we can find 10 different arrangements corresponding to 9 different blendings of the coffee of quality 1 and quality 2, bag for bag. We improve the quality of the first and we diminish the price of the second by blending.

Choose your coffees and prices, make some blendings and express the final price as a % of the price of the more expensive one. We know that in the previous example the price of the cheaper coffee is $\frac{7}{9}$ or $77 \cdot 7\%$ of the price of the more expensive coffee, so the prices of all the blendings we could make, exchanging bag for bag, were between $77 \cdot 7\%$ and 100% of the more expensive one.

If we wanted a blending to reach a price of 60% of that of the more expensive one, what should the price of the cheaper one be in relation to the more expensive one?

The answer to this question is as follows: if the cheaper coffee costs less than 60% of the other, we can make blendings with the other and reach that price. But if it were equal to or higher than 60%, that would be impossible. Is this true?

13.   Instead of coffee we could mix wines or other liquids, or we could form alloys that are mixtures of metals. The mathematical situation in any case would remain the same. The questions we could ask would concern the quality of the wine if the mixture is such and such; or the price at which the mixture is sold in order to yield such a profit; or a comparison of profits for various sales.

By making such problems yourself, you will understand much better all that is involved in these situations.

75

VII

PROBLEMS DEPENDING ON MEASUREMENTS
PROBLEMS IN ECONOMICS

Part VII

PROBLEMS DEPENDING ON MEASUREMENTS
PROBLEMS IN ECONOMICS

1. Most problems encountered in our daily life involve measurement (and what we have already learnt in this book will help us). Through the use of these measurements we can arrive at conclusions which may help us to decide what action to take. For instance, in building a house, or a flat, or buying the materials required to make instruments, or furniture, or a well, etc., measurements play an important role.

In Book II we learned to analyse prices in a manner that was rather oversimplified, but not altogether wrong. In this section we shall study more complicated situations and see how what we learnt in Book II and in this book can be used.

Building a road

2. Before we study the solutions we shall consider a complex problem and try to see what is wanted of us in order to solve it.

Let us say that we are asked to prepare a detailed budget for the construction of a road. There are a number of things that we must take into account:

(a) should the road serve for heavy or light traffic,

(b) should the road go through such and such a town or by-pass it,

(c) are bridges needed to cross rivers,

(d) are tunnels under railways, or through hills or mountains needed,

(e) should the road be of such quality that its maintenance would not involve continuous interruption of traffic, etc.?

Can you think of any other conditions that have to be put to the contractors who undertake the construction?

The contractors must, of course, know many things before they can submit a reasonable price in their tender, e.g.:

(1) the length of the road, and its required width;

(2) the quality of the soil: (are there swamps to drain? are there water infiltrations?—will the soil open up under pressure?—what differences in soil resistance are found in dry or wet weather?);

(3) how far must we go to find the materials needed to make the various layers of the road;

(4) the cost of materials, their transport, their treatment on the spot;

(5) the cost of labour and what goes with it: insurance, transport, overtime;

(6) the machinery needed; its maintenance, repairs, cost of replacement, deterioration;

(7) the risks of miscalculations and of the effect of unforeseen elements: an earthquake, a cloudburst, or only a quarrel with a sub-contractor, bringing a delay. Such factors are usually measured as a percentage of the cost.

Can you think of any other item that it is important to know? Among these items some are known after measurements and calculations and we shall consider these in particular.

3. A straight stretch of road is 4 km. long. It must first be cleared, flattened, rolled and compressed. It must be covered with layers of crushed stones, rolled and compressed with sand, then again covered by a layer of cemented sand and finally by a layer of tar.

If the road is 6 m. wide, can we find the amounts of materials needed, the number of men that have to work on it and the man-hours required, the cost of materials and labour, the time it will take to be completed, and the total cost?

All these calculations will only require multiplications and additions.

It is not in these arithmetical operations that we are really interested, but in the actual problems we have to solve before we can say for certain that the road will cost so much and it will take so long to build.

Find for yourself the information required in order to be able to answer the questions above in the case of the road 4 km. long and 6 m. wide which is supposed to be straight and built in successive layers as we have already outlined.

Is it sufficient to know the cost per km.?

Which of the factors involved here can be altered so that the time is shortened? Do you think that we can multiply the number of workers you suggest by 1,000 and reduce the time to $\frac{1}{1000}$ of its value? or have ten times more machinery?

Which variation of these factors could make the cost per km. higher in the case of a 4 km. road but may make it much lower for one of 40 km.? Can you see why?

Find the actual cost of road building in your country and try to analyse it so that it makes sense. Find actual costs in deserts, mountainous country, through thick forests, and compare them so that road building becomes for you a real problem.

Building site

4. Start with the plan of a building site. Choose a number of houses, the plan of each, the roads that link them, the area reserved for gardens for each house, for lawns on the site, the roundabout, the children's playground, etc.

From there, attempt to estimate the total cost of building the site, finding:

(a)   the cost of the land;

(b)   the cost of surveying and tracing the plans on the site;

(c)   the cost of the architect's share in preparing drawings for the projects;

(d)   the cost of the materials: sand, cement, water, bricks, tiles, window frames, glass panels, plumbing, wiring, paints . . .

80

(e) the cost of labour per man (bricklayers, plumbers, carpenters, painters . . .) per house, for the whole site;

(f) the transport of materials;

(g) the time taken to prepare the site, to build each house, and to make it ready for occupation.

Have we thus taken into account all that is really needed? Assuming a company builds the site, what should the selling price be, so that it can make a profit equal to a *fair* share of the investment? For that, think of what the money involved would have produced at 10% during the time taken, if it were, for example, loaned to an industrial concern.

Is it more profitable for the company to rent the houses or to sell them? What does the word 'profitable', cover? Is it more profitable for a rich man or for a medium income man, to receive rent instead of having his capital back?

5. Having spent so much for building a site and receiving so much in rents, it is easy to find what proportion of the capital the rent represents. Consider investments of money

(a) in savings banks;

(b) in stocks and bonds (through Stock Exchange);

(c) in industrial shares (through Stock Exchange);

and find out whether it is more profitable to invest in any of them. Can you measure the risks, in each, of not receiving the amounts calculated?

For example, from your newspaper, can you see how the news affects the value of shares, stocks, bonds, etc.?

Dividend and interest

6. *Interest* is the amount of money one receives when a given sum is loaned or invested in savings accounts.

Dividend is the amount profit-making concerns give to their shareholders.

When one borrows money, the lender demands a compensation for losing, for a while, the free use of his capital and for the help that capital will be to the new user. So it is an agreed practice that one pays interest on borrowed money.

81

When money is short and more people want it for their own use, the *rate of interest* or interest on a fixed sum (normally 100 units of money) goes up. When, on the contrary, money is plentiful, the rate goes down.

If *r* is the interest on £100, a simple proportion gives us the interest on any amount. Calculate the interest on a certain number of sums you choose, at a rate that is practised normally.

It is clear that if I use capital loaned to me for one or more years, the lender will feel, in ways that are not always identical, that I am depriving him of its use. So it is an agreed practice that interest should be paid at fixed intervals of time, usually a year or a fraction of a year. If it is for one year the interest is said to be *per annum*. For other periods a single proportion gives the answer. Find the formula for the interest on £100 at the rate *r* for (a) one year, (b) one month, (c) one week, (d) one day, (e) two years, (f) three and a half years.

If for a loan of £100 for 3 days, 10 units are demanded, what is the rate of interest? Do you think such a transaction is reasonable? Find in your dictionary the meaning of the word 'usury'. Could we apply it to this situation?

If the amount loaned or borrowed is a multiple of £100, another simple proportion will allow us to find the interest at a given rate for that sum. Choose:

sums as you wish

rates of interests as you wish (be reasonable)

durations of the loans as you wish

and find the amounts you would have to pay for the interest on those sums at those rates for those durations.

There are several notions that are useful in financial transactions. We shall mention their name here. Perhaps you will wish to know more about them. They are: compound interest; hire purchase; mortgage and insurance.

7.   From your newspaper you can gain information about industrial dividends. Can you find out:

(a) whether there exists a fixed rate of dividend?

(b) whether there are shares that have a better yield than others?

(c) what is a 'safe' investment?

If you had £1,000 at your disposal, which shares would you buy from those mentioned in the financial page of your newspaper? Justify your choices by calculations.

In your more specialised studies later on, you will learn much more about commercial and business transactions and the mathematics that is useful there. The ideas are many, the calculations only require additions, subtractions, proportions, and some other easy operations you will meet in Book VII, and later on.

Budgets

8. *Budgets* are calculations in advance of what amount of money may be needed for special purposes. We can have:

(a) individual budgets calculated for one week, one month, one year. Make your own budget for these three durations involving your pocket money, your transport fares, the cost of your food, your clothes, their cleaning, mending, etc. Think of other individual budgets, and work some out.

(b) family budgets involving rent, electricity and heating, cost of food, maintenance, outings, etc. Make a comparison of family budgets in various places. What are the differences in *standard of living* in one country among the various social and occupational groups; among various countries?

(c) Find out how sport clubs, charities, religious organisations, etc. get their money and run their finances. Study actual cases in your environment, and prepare a few society budgets.

(d) Most companies publish their balance sheets once a year. See whether you can find some in a newspaper. Compare the figures for income and expenditure, and for assets and liabilities, with the corresponding figures for the previous year, or for other companies.

See how much balance sheets can help in preparing a budget for the next year.

(e) Budgets of nationalised concerns and Government departments are published from time to time and tell how their revenue is intended to be spent. These show expenditures of different kinds:

(i) staff salaries and superannuation,

(ii) buildings to be erected and maintained,

(iii) projects to be carried out,

(iv) debts to be paid or interests on loans,

(v) deposits or advances to be paid, etc.

The revenue is usually calculated from the data of the previous year or as an amount allocated by the Government.

(f) National budgets include:

revenues from

(i) direct taxation (what is this?)

(ii) indirect taxation (what is this?)

(iii) nationalised industries and services

and expenditure on

(i) civil service

(ii) armed forces and police

(iii) armaments

(iv) cost of education (or the Government's contribution to this)

(v) health services

(vi) subsidies to government sponsored projects, etc.

Compare National budgets in different countries and find the percentage of expenditure that goes to the major items as well as the amounts that come from special services (such as postal services) and different taxes.

Consumer goods

9. The needs of individuals, of groups, of peoples, sometimes have numerical aspects.

For instance, *food* can be considered from that angle. If every member of a family takes coffee or tea three times a day, and uses two lumps of sugar each time, how much is needed per person, per family. Find out what would be needed by the inhabitants of various communities or countries.

Find, from the study of your own diet, approximately how much of the main items of food will be needed by your village, town or country per day and per year, assuming that all the people in your village eat the same quantity of food as you do, which is of course not the case.

Consider the consumption of tobacco, spirits, tea, bread and find approximately what is spent per year in your country on each item by the whole population.

Does expenditure on clothing follow the same pattern as the one on tobacco or tea?

Consider the various items of clothing and the average time they seem to last with you. Can you get an idea of the amount of business done by textile manufacturers, done by the manufacturers of ready made tailoring, underwear, footwear, in your country?

Travel

10. Car manufacturers have a certain production per year. Each car requires tyres, lining, glass, steel and instruments. Make a list of what is essential in cars and what spare parts are needed. Find out roughly how much material and how many different businesses are needed for the number of cars on the roads of your country.

Make a list of the useful but not essential parts in cars, and of the ornaments that you find in some cars.

As you see, one industry (here cars) creates and uses many others. Make a list of the manufacturers that are related to the manufacture of cars.

Choose the mileage done by a car in a day and calculate how much petrol it will consume in a year.

How much petrol will be needed for all the cars in your country, per day, per month, per year?

11. Air traffic is steadily increasing. Find out the consumption of petrol by the various types of aircraft and the amount of petrol needed by the air fleets flying on the main commercial and passenger routes. Find the daily and the annual consumption. Are these two numbers related in the proportion of 1 to 365? why not?

12. Problems depending on measurements are met in all aspects of life. They are not all of the same kind. But you will find that what you have learnt so far will very often prove useful in your daily life, and that with the few operations you know, you can work out many of the problems you may meet.

If you improvise your own problems and work on them, you will realise that a great variety of real life situations are related to a small number of mathematical solutions.

In Book VII new methods will be given that help in solving new types of problems. There are others that you will meet further on in your studies. New methods are invented all the time. You increase your powers by becoming acquainted with them.

VIII

PROBLEMS ON SPEED

PART VIII

PROBLEMS ON SPEED

Speed

1. If you walk or run steadily you will cover a certain distance in each unit of time, and we say that your *speed* is this distance per unit time. Whenever we know the distance covered and the time taken, we can calculate the speed by dividing the one by the other. For instance, 100 yards run in 10 seconds gives a speed of 10 yards per second. We can express the same result in different units, by using simple proportion. Thus if 100 yards is run in 10 seconds, then, if we assume the speed to remain unaltered, 600 yards would be run in one minute, and 36,000 yards in one hour. Since one mile is 1760 yards, we see that a 100 yards in 10 seconds corresponds to a speed of just over 20 miles per hour.

We know of course that even the best runners cannot run at this speed for a whole hour. But we can still say that they covered 100 yards at a speed of over 20 m.p.h. If we consider in more detail how 100 yards would be run in 10 seconds, we would find that the distance covered in the first second is less than 10 yards, because the runner is starting from rest. At the beginning his speed is increasing: this is called *accelerated* motion or acceleration. After reaching his top speed he will probably run at constant speed to the end of the 100 yards; any motion at constant speed in a straight line is called *uniform motion*. Finally, after passing the 100 yard mark, the runner will slow down; this slowing down is called *deceleration*. We could imagine that the speed at any point (being measured as the distance covered in a very short time-interval, perhaps much shorter even than a second) tells us something about the complete 'journey' of 100 yards and in this speed is usually referred to as the *average speed*.

Similar problems arise with cars. Find out the time taken for various car journeys, and calculate the average speed for these journeys. How long does it take to drive 5 miles at a speed of 30 m.p.h.? How much time would be saved by going at 40 m.p.h. instead? Why is the average speed for some journeys much higher than for others?

Paths

2. If you walk or run as you usually do when you go from one place to another, you rarely move along a straight line. Let us take 'A' and 'B' as the name of two places. Would there not be any number of *paths* that could join them on the ground? Imagine some of them. Which is the shortest?

Drawn on a piece of paper, these paths are *plane* paths, which they cannot be when you consider your own displacements. Plane curves or straight lines joining 'A' and 'B' can *represent* paths. Here are a few put together in one diagram:

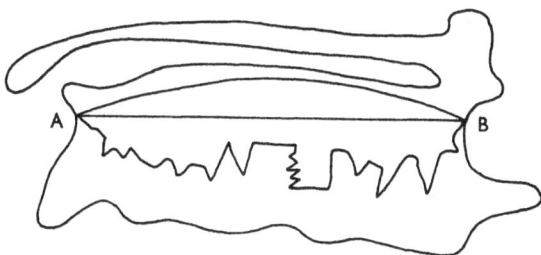

What could you say about them?

Speeds in various units

3. Lengths of paths are measured in units of distance. They may have, as we already know, several names according to the units of measurement. The names are equivalent if they represent the same length.

Similarly, time is measured by various units and the same duration may have different names.

It follows that speeds will have different equivalent expressions if various units are used to measure the distance between 'A' and 'B', or to measure the time taken to go from 'A' to 'B'.

For example, if AB equals 10 yds., or 30 ft. or 360 in., and if you take three seconds to go from 'A' to 'B', your speed will have three expressions: $\frac{10}{3}$ yds./sec. (read as $3\frac{1}{3}$ yds. per second), or 10 ft./sec. (read as 10 ft. per second), or 120 in./sec. (read as 120 in. per second). We know that variation of the unit of time would give us still other expressions.

If AB were equal to 100 miles, or 176,000 yds., or 528,000 ft., or 6,336,000 in., and if it took a car 3 hours, or 180 minutes, or 10,800 seconds to cover that distance, find the various equivalent expressions for the speed and enter them in the table:

	Miles	Yards	Feet	Inches
per hour	$33\frac{1}{3}$			
per minute				
per second				

Imagine different distances between two places 'A' and 'B' and different lengths of time taken by different means of communication to go from 'A' to 'B' (or 'B' to 'A'), and form a table giving the *various expressions* for the various speeds of the various means of communication.

Comparing actual speeds

4. Using the same units when possible and reasonable for the distances covered and for the time used to cover these, find out the speeds of various moving bodies around us, listing the slowest speeds first in a new table:

    a red globule or corpuscle in the blood

    an amoeba on a glass under the microscope

    Olympic sprinters

    the record for walkers (up-to-date)

    speed of horses, including race-horses

    speed of various animals

speeds of cars
speed of cyclists
speed of light in various media
speed of sound in various media
speed of planets, including the Earth
speed of aircraft of various types
speed of rockets
speed of satellites

Try to gather some ideas as to how these speeds are found.

Relative measurements

5. These speeds you found by looking around, are those we have been able to observe or calculate. The formula itself for speed that we have used (i.e. distance divided by time in appropriate units) does not impose a limit. So we could say that speeds can either be infinite or as small as we wished. A speed of zero is thought of as corresponding to a body at a standstill with respect to the system in which measurements are made. Immobility with respect to a body of reference is sometimes considered to be a motion of *speed zero* with respect to that body.

Thus, if a fly moves in a room, we can see that it covers a certain distance on a path during a certain length of time, and we can attach to its motion a certain speed with respect to, say the walls of the room. If it stays put on a wall for a while, its speed will then be zero with respect to the walls of the room. But the same fly in a car moving on the road will have speed zero with respect to the car if it stays put on a window, but will still have the speed of the car with respect to the road. Speed is a relative notion, relative to the system to which the particle or body considered is referred to.

Can you imagine when a fly in a car has a zero speed with respect to the road?

Imagine that the body of reference in which you study motion can be: (1) the car (moving or not moving), (2) the road or the earth, (3) the sun, (4) our galaxy, (5) another galaxy.

91

What can you say about the speed of a fly, or of a plane, or of a rocket, with respect to each of these bodies of reference?

Examples of problems on speed

6. Speed also has direction, for you can, on the same straight line or path, move from A to B or B to A. If you imagine two railway tracks on which trains run (without considering that they may run into each other), you can conceive of two trains having the same or different speeds, moving in the same or opposite directions, stopping sometimes or not stopping at all, leaving at the same time or at different times—and you can ask yourself various questions about them. This is the way some of the following questions have been invented. By changing the data you can of course obtain as many examples as you wish.

(i) Two trains leave town A to reach town B which is 20 miles from A. One is a goods train with a speed of 45 m.p.h., and the other is an express with a speed of 60 m.p.h. If the goods train leaves first, when would the express train be allowed to start on the same track if it were to reach B after the other train had arrived there?

(ii) Two towns, A and B, 70 miles apart, are linked by a railway. One train leaves A at 7 a.m. and its speed is 50 m.p.h. Another train, whose speed is 40 m.p.h., leaves B at 7.30 a.m. Where and at what time will they pass each other?

(iii) Two cities, A and B, are 450 miles apart and are linked by several trains a day. Between A and B there are other cities at which some of the trains stop and some do not. The stations in these cities have additional tracks, so that a non-stop train can overtake any one that is stationary there. If the speeds of these trains are 70 m.p.h., or 60 m.p.h., and if some of them stop sometimes for one minute, sometimes for up to five minutes at the intermediate cities, you can imagine that you may have to work out a time-table carefully to avoid accidents and unnecessary delays when slow trains are in front of faster ones between stations.

Start in the direction from A to B (450 miles apart); consider that one train leaves A every hour between 6 a.m. and

92

12 noon. Find, according to your choice of trains and of numbers of stops between A and B, what is happening on the tracks at 12.30 p.m. (i.e. find where the six trains you chose are, and the distances between them, numbering the trains 1, 2, 3 . . . 6 as they leave A). You may choose non-stop fast trains, fast trains stopping once for so many minutes or more than once for different lengths of time, or you can choose slower trains stopping or not stopping. Let us *not* take into account here that speed increases gradually when trains start. If we did that our questions would become much more complicated, as you will see later on.

So far we have tried to find out occurrences on the tracks with distances and speeds given. We can also find distances between cities, or speeds of trains, if we have the information that trains overtake each other or pass each other at a certain time or a certain place.

If two non-stop trains leave A at the same time on two tracks, and one reaches B two hours before the other, the speed of the faster train being 20 m.p.h. greater than that of the other, what else do you need to know in order to be able to find: (a) the distance from A to B, (b) the actual speed of the trains?

If two non-stop trains left, one A and the other B, at the same time and moved with the same speed, where would they meet? Can you say when? What else do you need to know in order to be able to answer this last question?

If one of these two trains was to stop for ten minutes at a place between A and a city just half-way between A and B, where would they meet? What else do you need to know in order to answer this question? Would it make any difference if both stopped, each for ten minutes before they met? Or after?

Do you think that the numbers (e.g. 450, 20 and 10) and the times that we have used in these problems for describing the distance between the cities, the difference in speed, and the length of stops and hours of arrival or departure can be changed without altering *the way* in which you handle the

93

situations in order to find the answers? If so, can't you ask or compose any number of questions of the same type?

What matters in these problems is that different patterns can be offered by the same situation, and it is for us to see how much can be changed, and in what way. For instance, if we wanted to find the distance between A and B, using only the data about the trains, would that be possible?

(iv)   Two trains whose speeds are 50 and 70 m.p.h. leave A to go to B. The faster one leaves one hour after the first. Both run non-stop and reach B at the same time. What is the distance between A and B? This is an example showing that we *can* answer the question above in the affirmative.

The solution is easy. We know that the second train will take as much time to catch up the 50 miles already travelled by the first train as the first one will take to reach B after having already been on the run for one hour. In other words, the difference of 20 miles per hour in the speeds of the two trains will serve, in $2\frac{1}{2}$ hours, to bring them level. This is sufficient for us to conclude that the first train has been on the run for $3\frac{1}{2}$ hours, and the second $2\frac{1}{2}$ hours. So we could say that AB=$3\frac{1}{2}\times 50$ miles or $2\frac{1}{2}\times 70$ miles=175 miles.

(v)   If we know that the distance between A and B is 600 miles, and that trains, whose difference in speed is 20 miles per hour, meet after 5 hours if they start at the same time, one from A and one from B, can we find the speeds of the two trains? (You can solve this problem with your rods, for if six orange rods end to end represent 600 miles, and in five hours one train does 100 miles more than the other, we can cover one orange rod of that length. There are now 500 miles left, or 250 miles to be covered by each of the two trains in 5 hours. This gives us the speed 50 m.p.h., which is the speed of the slower train. The speed of the faster one is 50+20=70 m.p.h.) Change the data and solve the new problems you obtain by doing so.

There can be many more variations on this theme of the trains and speeds. Improvise some and put the questions to your schoolmates to check whether you have taken into account the facts and not left out necessary data, or have put more information than required into your problems. Consult your friends also to check your solutions.

Graphical representation

8.   There is another way of representing the movements of trains on tracks. The various elements that we have considered so far, distances between cities, length of stops, speeds of trains, directions in which they move, can be shown graphically if we use squared paper to draw the graphs.

We can represent distances on one line as follows

A	B	C		D	E

and choose a unit of measurement that tells us that the distance between A and B is 30 miles, between B and C, 17 miles, between C and D, 32 miles, etc. Now, if we wanted also to show timings on the graph, we could take a perpendicular line, as in the following diagram, with the timings marked according to a unit of measurement. You will see that this graph is sufficient to show *all* we want, except tracks, which will require perhaps the use of colours.

This diagram tells the following story which you can follow step by step on it:

A goods train leaves A at 6 a.m. and reaches B 30 miles away, non-stop at 7 a.m. It stops there for 15 minutes and then proceeds to C 17 miles away, where it reaches at 7.50

95

a.m., stopping there for 20 minutes before returning non-stop to A, where it reaches at 9.40 a.m.

On this diagram you can find the speed of this train. It runs at the speed of 30 m.p.h. between A and B since it did 30 miles in one hour. The 17 miles from B to C were covered in 35 minutes, so the speed here is $\frac{17 \times 60}{35}$, that is, almost 30 m.p.h. since 17/35 is almost equivalent to $\frac{1}{2}$. The return speed is found by dividing 47 (miles) by 90 (minutes) which is a little more than 30 miles per hour (more exactly 90 min. = $1\frac{1}{2}$ hr., $47 \div 1\frac{1}{2} = 94 \div 3 = 31\frac{1}{3}$; speed $31\frac{1}{3}$ m.p.h.). Collect data from your nearest railway station about trains passing through your local station, stopping or not, and draw a diagram like the one above for some of the trains scheduled, including the return journey and all the stops, and find the speeds of your trains between stations. (Neglect as we did before, the time needed for slowing down before stopping, and for acceleration up to normal speed.)

If you use a few colours to represent different trains moving on the tracks during a certain length of time, you will get a graphical representation of what is happening on the railway, and like all the people whose function is the control of the trains on the tracks, you will be able to see at a glance if all the trains are running as they should, and are not in danger of colliding.

When tracings on your diagrams meet, that means that two trains are on the same spot at the same time. For collision to be avoided we shall have to place one train on one track and have another track free for the second train to pass that spot. (What are the main possibilities you can think of in avoiding collisions? List them.)

Such diagrams provide solutions to many problems that are hard to think about or to explain in writing even when using numerals. Here are a few examples for you to try out:

(1) It is known that, between A and E, distance 130 miles there is only one track, but that at B, C, D and E, there are several tracks at the stations where trains can stop to let others overtake them. Choose speeds of trains, times of

departure from A and from F as well as distances between stops and find out whether the trains can all be on the track safely, travelling in both directions, and safely reach their destinations (A or F), using suitable stops on the way. See whether any trains are causing danger, or making traffic too difficult, and, if so, remove them. To see whether you have done it well, consider carefully whether traffic is now really safe.

(2)   We now have two tracks that we can use, but each only in one direction. Consider the same questions as in the previous problem.

(3)   The two tracks can be used in both directions, thanks to safety devices that close a track to any new train, if it is already carrying another. This is a complication for you but a much more efficient use of the tracks. Do you see why? Can you solve the problems set in (1) on such a railway system?

Acceleration

9.   In all we did above, we considered only uniform speeds. But we know in our experience that speed can change when a vehicle is in motion.

Acceleration is the increase of speed per unit of time. Similarly a reduction of speed is called deceleration and this may also be expressed as a negative acceleration. All these expressions are in use, and it is worth knowing them.

Acceleration however, does not only result from a change in the magnitude of the speed (e.g. from 30 m.p.h. to 20 m.p.h. or to 40 m.p.h.), but it can also be caused by a change in the shape of the path from a straight line to a curve. When you travel in a car or train you can always feel that on a bend, if the vehicle does not slow down, you are pushed out of your seat. It is the effect of the acceleration due to the change of direction of the moving body; change that can be said to be of the direction of the speed of the vehicle.

Because speed has magnitude and direction, it is an example of a *vector* quantity, and when we think of both these qualities, it is more usual to speak of *velocity*, leaving the word speed to the cases when we do not need to know that a

97

direction is involved. When we use the word speed we think of a ratio representing a magnitude (so many units of space covered during so many units of time). When we use the word velocity we see associated with the moving object an arrow indicating a direction as well. This you will understand better in your studies of physics.

IX

THE PERPETUAL CALENDAR

Part IX

THE PERPETUAL CALENDAR

Use of remainders in division by 7

1. In Book II we used the rods to establish more clearly the structure of the Gregorian Calendar, and we learnt to find the day of the week for any date within one year. In this section we are going to investigate the structure of the calendar a little further so as to be able to find the day of the week of any date.

2. We can number the days of the year consecutively, counting the 1st January as No. 1, the 1st February as No. 32, and so on, until we reach 31st December which will be No. 365 in an ordinary year and No. 366 in a leap year.

When you divide any number by 7, the remainder may be either 1, 2, 3, 4, 5, 6 or the number may divide exactly without remainder, in which case we can call the remainder 0. You will see that in any year there is a relation between the day of the week and the remainder we get when dividing the number of the day by 7. The reason for this is that the remainders repeat themselves after 7 days, and so do the days of the week.

Make a table of all the dates this year which have the same day of the week as today, but *do not* consult a calendar. In which months do you get 5 of these days?

Days in the year

3. If we know what day of the week the 1st January is, we can easily calculate the day for any other date in the same year. Take, for instance, the 5th April. In an ordinary year this is the 95th day of the year. Dividing by 7 we get the remainder 4, so that April 5th will fall on the same day of the week as 4th January. What happens in a leap year?

100

Our answers become particularly simple if the 1st January is a Monday, which we also take as the first day of the week. Then the 4th January would be a Thursday, and (if it is not a leap year) the 5th April would also be a Thursday.

It so happens that in the year in which this book was first published, 1962, the 1st January was a Monday.

The number code for the month

4. Is it necessary to give each date its number in the year? Since January has 31 days, it has three days in excess of four weeks, and so the 1st February is the same day of the week as the 4th January. Similarly the 14th February is the same day as the 17th January (and the 3rd January).

By writing down for each month its excess of days over 28, and adding these numbers, you obtain a number corresponding to the shift in days of the week for the following month. Taking January as the starting month, we give February the number 3, March the same number 3, and April the number 6 (in an ordinary year, of course). Instead of giving May the number 8, we can give it the number 1, since we are only interested in the remainders of division by 7. Continuing in this way you should get the number 5 for December.

As an example we shall now consider the date 23rd April (St. George's Day). Add the code-number 6 for April, and we get $23+6=29$; remainder after division by 7 is 1, hence the day of the week is the same as for 1st January (in an ordinary year). Hence the 23rd April, 1962 was a Monday.

Days in different years

5. We shall now consider how the days of the week shift from one year to the next. An ordinary year has 365 days, which gives 52 weeks and one day over. So from one year to the next there is a shift of one day, except in leap years where there is a shift of two days.

From 1900 to 1962 we have 62 years, of which 15 were leap years. (Remember that in our calendar 1900 was not a leap year). Since $62+15=77$, and 77 is divisible by 7, we see that the calendar for 1900 would have shown the same days of the week as for 1962. This shows us that 1st January, 1900 was a Monday, and since 1900 was *not* a leap year, we shall find this date a useful basis for calculating the day of the week of any date.

Some historical dates worked out

6. For historical dates of the past we have to subtract instead of adding the shifts. We must also remember that in our calendar 1900 was not a leap year. If we go further back than the nineteenth century we must take into account the days lost during the change of calendar which took place in 1582 in most countries on the Continent, in 1752 in Great Britain, and still more recently in Russia and Greece. Some countries are still using the old calendar.

Here are some examples of historical dates:

(a) The first world war started on 2nd August, 1914: From 1900 to 1914 there are 14 years, of which 3 are leap years, giving a shift of 3 days. 1914 was not a leap year, so the code for August is 2. Add 2 for the date 2nd August, and we get $3+2+2=7$. Since the first day in this calculation was a Monday (1st January, 1900), the day corresponding to a 7 is a Sunday.

(b) The second world war started officially on 3rd September, 1939: From 1900 to 1939 there are 39 years, of which 9 were leap years. The code for September in an ordinary year is 5. Now $39+9+5+3=56$; divide by 7, and the remainder is 0, which again corresponds to a Sunday. Since we are only interested in the remainders, we could have replaced each of the numbers 39 and 9, by its remainder, getting $4+2+5+3=14$, and again the remainder is 0.

(c) The first atomic bomb was dropped on Hiroshima on 6th August, 1945: We get $45+11+2+6$, remainder 1, hence this was a Monday.

(d) The storming of the Bastille (French Revolution) on 14th July, 1789: From 1789 to 1900 is 111 years, including 26 leap years (1800 and 1900 were *not* leap years). $111+26$ divided by 7 leaves remainder 4. Now 14th July, 1900 has the same remainder as the code for July, namely 6, and to go *back* to 1789 we have to *subtract* 4, giving 2. It was a Tuesday.

Make up your own examples from the date of birth and birthdays of your friends.

Find the day of birth of great men whose date of birth you can find in an historical dictionary. Here are a few names.

Bach	Dante	Newton
Beethoven	Cervantes	Laplace
Mozart	Shakespeare	Gauss
Chopin	Milton	Faraday
Tchaikovsky	Goethe	Maxwell
Debussy	Tolstoi	Hertz

You can increase that list indefinitely and work on it to master the perpetual calendar, and get fun from using it.

7. *Historical note on the calendars*

The Ancient Egyptians used a solar year of 365 days related to agriculture and the cycle of the seasons. Every four years there was a shift of one day, and after 15 centuries the seasons' beginning returned to the same date, having gone through all days of the year. Not all calendars are solar; some are lunar, as is the case with the Moslem calendar which counts a year of 12 lunar months and its year lasts for 354 or 355 days.

The Jewish year counts 12 or 13 lunar months: two years of 12 months lasting 353, 354 or 355 days are followed by one of 13 months that count 383, 384, 385 days. This type of year is called luni-solar. The Hindu people also use a similar calendar.

The Coptic calendar was less adapted to seasons than the Egyptian, and Julius Caesar in 45 B.C. reformed it according to the data provided by the Alexandrian astronomer Sosigenes. The year was equal to 365 days and a quarter, with 3 ordinary years of 365 days, followed by one leap year of 366. The Romans doubled the 24th day of February, named by them the VIth day before the calends of March, (hence the continental name of bi-sextile for the leap year).

103

The Christian Church maintained the Julian calendar. But the Julian year loses one day every 128 years, and in 1582 Pope Gregory XIII decided to cut 10 days off that year to re-establish the relation between the calendar and the seasons to what it was at the time of the Council of Nicaea (325) when the feast of Easter was fixed. (In Rome the 4th of October 1582 was followed by the 15th, in France the cut was made in December, and in Great Britain it was only adopted in September, 1752, not without opposition from the people. In Russia the calendar was adopted in 1920). In the Gregorian calendar the century years whose dates are divisible by 400 are to be leap years, while the others are not. So 1700, 1800, 1900 were not leap years while 1600 was and 2000 will be.

The loss, in the Gregorian calendar, of a whole day, due to our failing to take into account all the astronomical facts, will be completed when we reach the years around 4500.

www.ingramcontent.com/pod-product-compliance
Lightning Source LLC
Chambersburg PA
CBHW060619200326
41521CB00007B/825